Toward Climate-Resilient Development in Nigeria

DIRECTIONS IN DEVELOPMENT
Countries and Regions

Toward Climate-Resilient Development in Nigeria

Raffaello Cervigni, Riccardo Valentini, and Monia Santini, Editors

THE WORLD BANK
Washington, D.C.

Contents

Maps

Tables

Foreword by Nigeria's Coordinating Minister for the Economy

Over the last decade, Nigeria has experienced steady growth, averaging over 7 percent per annum in the last five years. Even though these figures are healthy relative to several other countries, we need to make sure our economy grows more resilient to both short-term financial and socioeconomic shocks and longer-term stressors. We are making progress in strengthening our resilience to economic shocks, for example, by managing our foreign reserves and excess crude account balances; or by investing in sectors that can help diversify growth away from oil, such as agriculture, power, housing and construction, solid minerals, education, health, information and communication technology, and others.

In addition to protecting our growth from economic shocks, however, we need to protect it from climate-related ones. The 2012 floods were an abrupt reminder of the vulnerability of our communities and infrastructure to natural disasters. What the future has in store for us is more erratic weather, and with it, the risk of more frequent and more severe extreme events.

This book provides a comprehensive overview of the likely impacts of climate change on sectors that are strategic for the growth of our economy, such as agriculture, livestock, and water resource management. It alerts us that increases in temperature, coupled with changes in precipitation patterns and hydrological regimes, can only exacerbate existing vulnerabilities.

The book also highlights the fact that there are promising opportunities to build resilience into the fabric of our economy. We can start exploring those opportunities by focusing our efforts where they matter the most: for example, in agriculture, which contributes about 40 percent of our GDP, and employs more than half our workforce. President Goodluck Jonathan's Transformation Agenda provides a strategic platform to raise the sector's productivity, attract private sector investment, and reduce excessive dependence on food imports. Increasing our ability to respond to natural disasters—and eventually prevent their deleterious impacts—is another important area. The recent establishment of the Inter-Ministerial Committee on a National Framework for Application of Climate Services is an important step in this direction.

As we move toward the next stages of implementation of these initiatives, the material contained in this book will be particularly valuable, helping to inform decision making across sectors and levels of governments, and to ensure that the economy becomes not only more productive but also more climate resilient.

Ngozi Okonjo-Iweala
Coordinating Minister for the Economy and
Federal Minister of Finance

Foreword by the World Bank

Vision 20: 2020 sets out Nigeria's ambition to become one of the World's 20 largest economies by 2020. Climate change poses challenges to reaching this goal but also creates opportunities. In the aftermath of the Rio+20 Summit, concepts such as green growth, low-carbon development, and carbon-resilient economies are beginning to drive development policies and investments.

For the past two years, the Federal Government of Nigeria and the World Bank have collaborated to analyze the specific challenges posed by climate change in agriculture and water resources management. This effort has brought together participants from federal and state government agencies, academia, the private sector, civil society, and the community of development partners to discuss how the economy can be made more climate-resilient, in accordance with the National Adaptation Strategy and Plan of Action on Climate Change for Nigeria. The World Bank welcomes this partnership and recognizes its significance for Nigeria's development—and for the leadership role Nigeria can play in moving Africa forward in the global effort to respond to climate change.

Toward Climate-Resilient Development in Nigeria summarizes the final results of the analysis, with a sobering message on the climate change impacts that can be expected if timely actions are not taken. If not addressed, climate change will limit Nigeria's ability to achieve and sustain the goals set out in Vision 20: 2020. Fortunately, a range of technological and management approaches can enable the country to better handle current climate variability and build resilience to the harsher climate of the future. The book proposes methodological innovations such as the application of "robust decision making" to irrigation and hydropower development when the future climate is uncertain. These strengthen the case for immediate action.

The book proposes 10 practical short-term priority actions, as well as complementary longer-term initiatives, that could help to mitigate the threat to Vision 20: 2020 that climate change poses. Nigeria's vision can become a reality if the country moves promptly to become more climate-resilient. The World Bank is ready to support the Government in this effort, and looks forward to reinforcing its partnership with Nigeria's federal and state governments on climate action,

seizing opportunities for cross-sectoral investments and offering support for
policy reforms.

Marie Francoise Marie-Nelly   Jamal Saghir
Country Director for Nigeria   *Sector Director*
The World Bank   *Sustainable Development Department,*
                                                          Africa Region
                                                       The World Bank

Acknowledgments

This book was prepared by a World Bank team led by Raffaello Cervigni and including (in alphabetical order) Amos Abu, Abimbola A. Adubi, Joseph Ese Akpokodje, Benedicte Marie Cecile Augeard, Ademola Braimoh, Stephen Danyo, Irina Dvorak, Erik Magnus Fernstrom, Francesca Fusaro, Sarwat Hussain, Ella Omomene Iklaga, Anushika Karunaratne, Rikard Liden, Stephen Ling, Kiran D. Pandey, Beula Selvadurai, and Shobha Shetty.

The analysis is based on a consulting report prepared by a team of researchers of the Euro-Mediterranean Center on Climate Change (CMCC), led by Riccardo Valentini and including, in alphabetical order, Francesco Bosello, Jaroslav Mysiak, Monia Santini, Pasquale Schiano, Enrico Scoccimarro, and Donatella Spano. The extended CMCC team included Ahmed Balogun, Edoardo Bucchignani, Lorenza Campagnolo, Gianluca Carboni, Fabio Eboli, Chibueze Emenike, Andrea Gallo, David Onemayin Jimoh, Valentina Mereu, Myriam Montesarchio, Antonio Navarra, and Antonio Trabucco.

The World Bank Group peer reviewers were Erick Fernandes, Alan Miller, and Rita Cestti.

The work was undertaken under the supervision of the following World Bank managers: Idah Z. Pswarayi-Riddihough (Sector Manager, Environment, Natural Resources and Climate Change in Africa); Jamal Saghir (Director, Africa Sustainable Development Department); and Marie Francoise Marie-Nelly and Onno Ruhl (respectively, current and former Country Director for Nigeria).

The book was edited by Anne R. Grant (consultant); Kiran D. Pandey coordinated the inputs from the various authors to the final version of the book. In the World Bank's Office of the Publisher, Stephen Pazdan oversaw the book production process and Patricia Katayama handled acquisition duties.

From the Government of Nigeria, policy guidance was provided by the Hon. Mrs. Hadiza Ibrahim Mailafa, Federal Minister of Environment; the Hon. Dr. Akinwunmi Ayo Adesina, Federal Minister of Agriculture; and the Hon. Mrs. Sarah Ochekpe, Federal Minister of Water Resources.

The study team wishes to thank the following for their support and for access to data: Amb. Godsend Igali, Permanent Secretary, Federal Ministry of Water Resources; Dr. Jare Adejuwon, Director, Department for Climate Change, Federal Ministry of the Environment; Professor A. S. Sambo, Director General/ CEO, Energy Commission of Nigeria; Dr. A. C. Anuforom, Director General/

CEO Nigerian Meteorological Agency (NIMET); Mr. Gambo Jakanda, Acting Permanent Secretary, Federal Ministry of Environment; Eng. John A. Shamonda, Director General, Nigeria Hydrological Services Agency (NIHSA); Dr. Sanusi Garba, Director of Power, Minister of Energy; Mr. Joseph Alozie, Deputy General Manager, Climate Services, NIMET; Mr. Tunde Lawai, Acting Director, National Planning Commission; Eng. M. C. C. Eneh, Director, Federal Department of Agriculture; I. D. Nnodu, Acting Director, Nigerian Meteorological Agency; Eng. Razaq A. K. Jimoh, Coordinating Director, Nigeria Integrated Water Resources Management Commission; Mr. Simon Harry, Head, Project Implementation Task Team, National Bureau of Statistics; P. A. Nlekwa, Director, Hydrogeo-informatics, NIHSA; and Eng. Patrick Nlekwa, Director, Hydrogeography, NIHSA.

The team is also grateful for the comments received from participants to the final workshop held in Abuja in December 2012, including: Yakubu John Akubo (FMARD), Omotola Temitope (NPC), Oyewole Patrick (FMARD), Abah Philip (FMARD), Abioye Aderemi (FMARD), David O. Jimoh (Federal University of Minna), Sule Bolaji (University of Ilorin), Otun Johnson (Ahmadu Bello University, Zaria), Comfort A. Owolabi (Department of Climate Change, FME), Saico Comfort (PHCN PMU), Sa'adatu Gambo Madaki (Department of Climate Change, FME), Ololade Adegbite (FMARD), Uche Egwuatu (FME), Aishatu S. Ingawa (ECN), Jonah D. Barde (FME), Richard O. Nzekwu (FMARD), Toyin Ohigbemi (ECN), Eric Florimon-Reed (USAID, Nigeria), Michael B. Idowu (Abeokuta), S. M. Wilson (NIMET), Jasper Kemakolam (NIMET), Nathan Awuapila (ECN), Henry Agbnika (CIDA), Mrs. C. N. Njokwu (FMPR), A. Y. Anda (NIWRMC), C. O. Ezinma (FMARD), R. A. Adetowo (FME/Fadama), S. M. Babarinda (FMARD), Ngingifa-Williams (Federal Ministry of Works and Housing), Oladoyin Odubanjo (Nigerian Academy of Sciences), Dikson Okolo (FMARD), A. Adebisi (FMP), T. O. Dina (FMP), S. B. Ayangoaor (FMP), H. M. Gidado (ECN), H. S. Adamu (ECN), Ibrahim Umaru (Nasarawa State University, Keffi), P. B. Aribo (NSLM), Emenike Chibueze (Heinrich Boell Stiftung), Godsend Igali (Federal Ministry of Works and Rural Development), and J. A. Shamonda (NIHSA).

Financial support from the following trust funds is gratefully acknowledged: the Trust Fund for Environmentally and Socially Sustainable Development (TFESSD), the TerrAfrica Leveraging Fund, the Bank-Netherlands Partnership Program (BNPP), and the Energy Sector Management Assistance Program (ESMAP).

About the Editors

Raffaello Cervigni is a lead environmental economist with the Africa Region of the World Bank. He holds master's and PhD degrees in economics (from Oxford University and University College London), and has some 18 years' professional experience on programs, projects, and research financed by the World Bank, the Global Environment Facility, the European Union, and the Italian Government in a variety of sectors. He is currently the World Bank's regional coordinator for climate change in the Africa Region, after serving for about three years in a similar role for the Middle East and North Africa Region. He is the author or coauthor of more than 40 technical papers and publications, including books, book chapters, and articles in learned journals.

Riccardo Valentini is professor of forest ecology in the Department for Innovation in Biological, Agro-food, and Forest Systems of the University of Tuscia and director of the Division of Impacts on Agriculture, Forest, and Natural Ecosystems (IAFENT) at the Euro-Mediterranean Center on Climate Change (CMCC). His expertise concerns the role of climate, land use change, and forestry in the carbon cycle, biodiversity, and bioenergy. He is a member of the Intergovernmental Panel on Climate Change and has been coordinator of several EU projects. In 2010 he was awarded a European Research Council grant to lead a project on "The Role of African Tropical Forests in the Greenhouse Gases Balance of the Atmosphere." From 2009 to 2012 he chaired the Global Terrestrial Observing System. He has published more than 100 papers in international journals.

Monia Santini, who holds a PhD in forest ecology, is the scientific coordinator of the Impacts on Agriculture, Forest, and Natural Ecosystems (IAFENT) Division at the Euro-Mediterranean Center on Climate Change (CMCC). Her research interests are climate change and its interactions with land cover and water uses. She specializes in constructing and applying process-based models to account for the interactions between the climate and land surface components of the hydrologic cycle (vegetation, soil), to better reproduce and project the impact of climate on ecosystems and the services they provide.

Abbreviations

ADP	Agricultural Development Program, Agricultural Development Project
AESZ	agro-ecological subzone
AEZ	agro-ecological zone
AR4	IPCC Fourth Assessment Report
ARCN	Agricultural Research Council of Nigeria
ATA	Agricultural Transformation Agenda
BNRCC	Building Nigeria's Response to Climate Change
CCSR	Center for Climate System Research
CGCM	Coupled General Circulation Model
CGE	computable general equilibrium
CIDA	Canadian International Development Agency
CMCC	Euro-Mediterranean Center on Climate Change
CMIP3	Coupled Model Intercomparison Project (Phase 3)
CNRM	Centre National de Recherches Météorologiques
CO_2	carbon dioxide
COSMO-CLM	COSMO (Consortium for Small-Scale Modeling) Model in Climate Mode
CRA	climate risk analysis
CRU	Climate Research Unit
CSA	climate-smart agriculture
CSIRO	Commonwealth Scientific and Industrial Research Organization
DEM	Digital Elevation Model
DSSAT-CSM	Decision Support System for Agrotechnology Transfer—Cropping System Model
ECN	Energy Commission of Nigeria
ESO	Explicit Stochastic Optimization
FACE	free air carbon dioxide enrichment
FAO	Food and Agriculture Organization of the United Nations

FGN	Federal Government of Nigeria
FMARD	Federal Ministry of Agriculture and Rural Development
FME	Federal Ministry of the Environment
FMP	Federal Ministry of Power
FMPR	Federal Ministry of Petroleum Resources
GCM	General/Global Circulation Model
GDP	gross domestic product
GEF	Global Environment Facility
GFDL	Geophysical Fluid Dynamics Laboratory
GIS	Geographic Information System
GPP	gross primary productivity
GW	gigawatt
HA	hydrological area
ha	hectare
HDI	Human Development Index
HRU	hydrological response unit
IAP	Institute of Atmospheric Physics
ICASA	International Consortium for Agricultural Systems Applications
ICES	Intertemporal Computable Equilibrium System
ICID	International Commission on Irrigation and Drainage
IFPRI	International Food Policy Research Institute
IIASA	International Institute for Applied Systems Analysis
IMPACT	International Model for Policy Analysis of Agricultural Commodities and Trade
IPCC	Intergovernmental Panel on Climate Change
JICA	Japan International Cooperation Agency
kWh	kilowatt hour
MAR	mean adequacy ratio
MDA	ministries, departments, and agencies
MDGs	Millennium Development Goals
MPI	Max Planck Institute
MRI	Meteorological Research Institute
MT	metric ton
MW	megawatt
NAR	nutrient adequacy ratio
NARS	National Agricultural Research System
NASPA-CCN	National Adaptation Strategy and Plan of Action on Climate Change for Nigeria

NASRDA	National Space Research and Development Agency
NCAR	National Center for Atmospheric Research
NCCAS	National Climate Change Adaptation Strategy
NCCC	National Climate Change Commission
NEEDS	National Economic Empowerment and Development Strategy
NEWMAP	Nigeria Erosion and Watershed Management Project
NFDP	National Fadama Development Project
NGO	nongovernmental organization
NIHSA	Nigeria Hydrological Services Agency
NIMET	Nigerian Meteorological Agency
NIWRMC	Nigeria Integrated Water Resources Management Commission
NPC	National Planning Commission
NPV	net present value
NWRMP	National Water Resources Master Plan
O&M	operation and maintenance
PHCN	Power Holding Company of Nigeria
ppm	parts per million
R&D	research and development
RBDA	River Basin Development Authority
RCM	regional climate model
RDM	robust decision making
RNI	recommended nutrient intake
SAM	social accounting matrix
SCCU	Special Climate Change Unit (of FME)
SCGT	single-cycle gas turbine
SDAP	Sustainable Development Action Plan
SDP	Stochastic Dynamic Programming
SFM	sustainable forest management
SLM	sustainable land management
SPAM	Spatial Production Allocation Model
SSA	Sub-Saharan Africa
SWAT	Soil and Water Assessment Tool
SYC	storage-yield curve
THI	temperature-humidity index
TOR	terms of reference
UKMO	United Kingdom Meteorological Office
UNCBD	United Nations Convention on Biological Diversity
UNCCD	United Nations Convention to Combat Desertification

UNDP	United Nations Development Programme
UNFCCC	United Nations Framework Convention on Climate Change
URV	unit reference value
WHO	World Health Organization
WMO	World Meteorological Organization

Overview

Main Policy Messages

If not addressed in time, climate change is expected to exacerbate Nigeria's current vulnerability to weather swings and limit its ability to achieve and sustain the objectives of Vision 20: 2020 (as defined in http://www.npc.gov.ng/home/doc.aspx?mCatID=68253). The likely impacts include:

- A long-term reduction in crop yields of 20–30 percent
- Declining productivity of livestock, with adverse consequences on livelihoods
- Increase in food imports (up to 40 percent for rice long term)
- Worsening prospects for food security, particularly in the north and the southwest
- A long-term decline in GDP of up to 4.5 percent.

The impacts may be worse if the economy diversifies away from agriculture more slowly than Vision 20: 2020 anticipates, or if there is too little irrigation to counter the effects of rising temperatures on rain-fed yields.

Equally important, investment decisions made on the basis of historical climate may be wrong: projects ignoring climate change might turn out to be either under- or over-designed, with losses (in terms of excess capital costs or foregone revenues) of 20–40 percent of initial capital in the case of irrigation or hydropower.

Fortunately, there is a range of technological and management options that can be adopted both to better handle current climate variability and to build resilience against a harsher climate:

- By 2020 sustainable land management practices applied to 1 million hectares can offset most of the expected shorter-term yield decline; gradual extension of these practices to 50 percent of cropland, possibly combined with extra irrigation, can also counterbalance longer-term climate change impacts.
- Climate-smart planning and design of irrigation and hydropower can more than halve the risks and related costs of making the wrong investment decision.

---

box continues next page

Main Policy Messages *(continued)*

The Federal Government could consider 10 short-term priority responses to build resilience to both current climate variability and future change through actions to improve climate governance across sectors, research and extension in agriculture, and hydrometeorological systems; integration of climate factors into the design of irrigation and hydropower projects; and mainstreaming climate concerns into priority programs, such as the Agricultural Transformation Agenda.

Climate Risks to Vulnerable Sectors

This book analyzes the risks to Nigeria's development prospects that climate change poses to agriculture, livestock, and water management. These sectors were chosen because they are central to achieving the growth, livelihood, and environmental objectives of Vision 20: 2020; and because they are already vulnerable to current climate variability. Since other sectors might also be affected, the findings of this research provide lower-bound estimates of overall climate change impacts.

Agriculture accounts for about 40 percent of Nigeria's GDP and employs 70 percent of its people. Because virtually all production is rain-fed, agriculture is highly vulnerable to weather swings. Stagnating yields in the presence of a growing population are causing dependency on food imports (particularly rice) to increase. In large parts of the country, especially in the northern states, livelihoods depend on livestock, which accounts for 5 percent of GDP; livestock is already exposed to thermal stress, and to declining pasture productivity.

Climate variability is also undermining Nigeria's efforts to achieve energy security. Though dominated by thermal power, the country's energy mix is complemented by hydropower, which accounts for one-third of grid supply. Because dams are poorly maintained, current variability in rainfall results in power outages that affect both Nigeria's energy security and its growth potential.

Climate change is likely to make food, energy, and water security harder for Nigeria to achieve. Various climate modeling exercises all underscore the severity of the challenge resulting from temperatures expected to rise an average of 1–2°C by 2050 (figure O.1), and even more during the winter.

The quantity of water available for storage and use will change. Conflicting precipitation projections make it difficult to estimate how much water could ultimately be directed to irrigation, hydropower, and municipal water supplies, but a consensus is emerging that for close to 80 percent of the country, using past climate to guide design of future water management projects might lead to inappropriate investment decisions (map O.1).

In particular, climate models converge in projecting that by mid-century water flows will increase for almost half the country, decrease in 10 percent of the country, and be uncertain over one-third of Nigeria's surface. Stable conditions are projected only in 8 percent of the national territory.

Figure O.1 Average Air Surface Temperature, Nigeria, 1976–2065

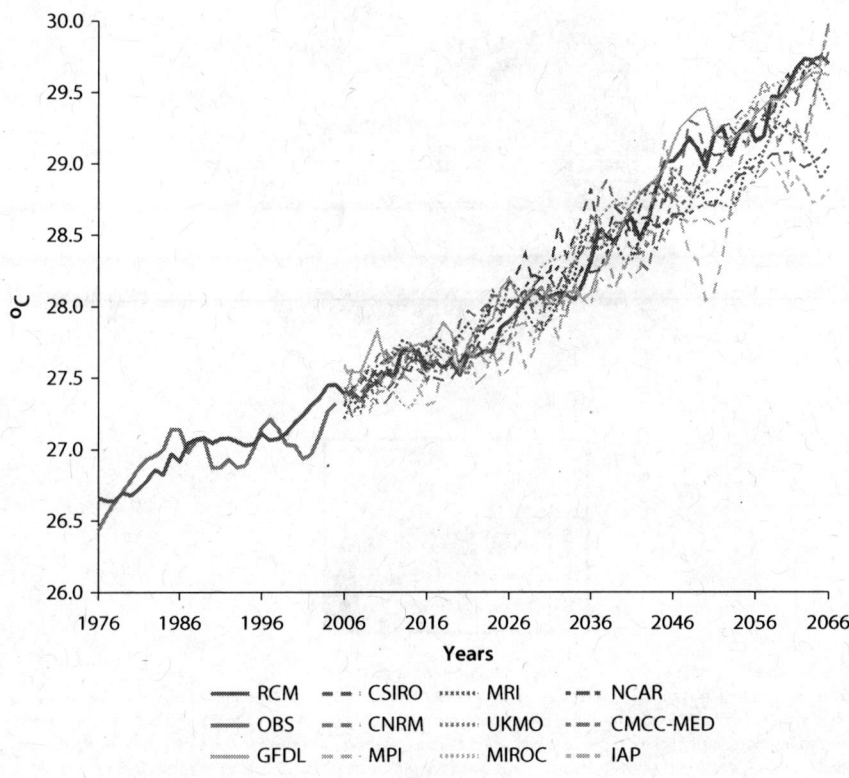

Source: Authors' calculations based on data sources listed in table 3A.1.
Note: The lines represent trended observation (solid orange line) and climate model simulations from 2006.
OBS = observation; RCM = Regional Climate Model; CMCC-MED = Euro-Mediterranean Center on Climate Change;
CNRM = Centre National de Recherches Météorologiques; CSIRO = Commonwealth Scientific and Industrial Research
Organization; GFDL = Geophysical Fluid Dynamics Laboratory; IAP = Institute of Atmospheric Physics; MIROC = Center for
Climate System Research; MPI = Max Planck Institute; MRI = Meteorological Research Institute; NCAR = National Center for
Atmospheric Research; UKMO = United Kingdom Meteorological Office.

A Decline in Rain-Fed Yields

Even if precipitation increases in several parts of the country, this is not likely to
offset the negative effects of rising temperatures on yields of most rain-fed crops,
particularly over the long term (figure O.2).

The shorter-term effects are more uncertain: by 2020, according to more than
half the climate models, yields for cassava and perhaps other crops might actually
increase.

Implications for GDP Growth and Trade

Climate-induced declines in crop yields are expected to have significant long-
term effects on the GDP of Nigeria, reducing it by as much as 4.5 percent by
2050. These modeling results assume—in accordance with Vision 20: 2020—
that the share of agriculture in GDP will decline from 40 to just 15 percent.

Map O.1 Projected Changes in Average Water Flows by Sub-basin, 2050 Compared to 1990

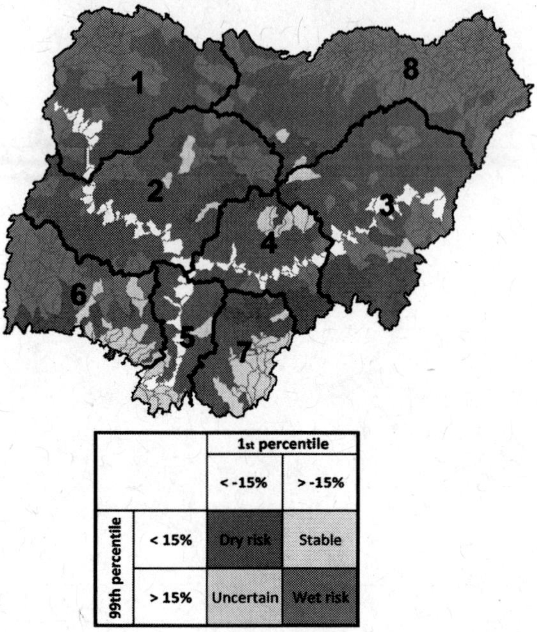

		1st percentile	
		< -15%	> -15%
99th percentile	< 15%	Dry risk	Stable
	> 15%	Uncertain	Wet risk

Source: Authors' calculations based on data sources listed in table 3A.1.
Note: For water planning purposes, a change of less than +/–15 percent in historical water flow is considered equivalent to stable conditions. Colors describe the consensus among climate models on whether the stability band is likely to be exceeded by mid-century. For example, blue areas indicate basins where most models agree that water flows will increase more than 15 percent; in red areas, the consensus is that flows will decrease by 15 percent or more. The numbers on the map refer to hydrological areas.

If the economy diversifies away from agriculture at a slower pace, the negative effects on GDP growth are likely to be much larger.

The analysis also assumes a sustained rate of expansion of irrigated area until it reaches its long-term potential of 11 million hectares (ha). Less irrigation will make agriculture less resilient to the decline of rain-fed yields, jeopardizing the goal of achieving food security.

Climate change is also projected to affect food trade. While in the long term net imports of yams and other vegetables could decrease, net imports of cereals are expected to increase, by as much as 40 percent in the case of rice, as demand increases but yields decline.

Challenges to Food Security

It is projected that by 2020 half of Nigeria's agro-ecological zones (AEZs) will not be able to meet demand for food with local supply; by 2050 75 percent will be in the same position. Food security is thus in danger unless the decline in local food production is offset by vast improvements of in-country trade in food grains and more food imports. In both cases major investments in transport and storage

Figure O.2 Aggregate Percent Change in Crop Yields by 2050

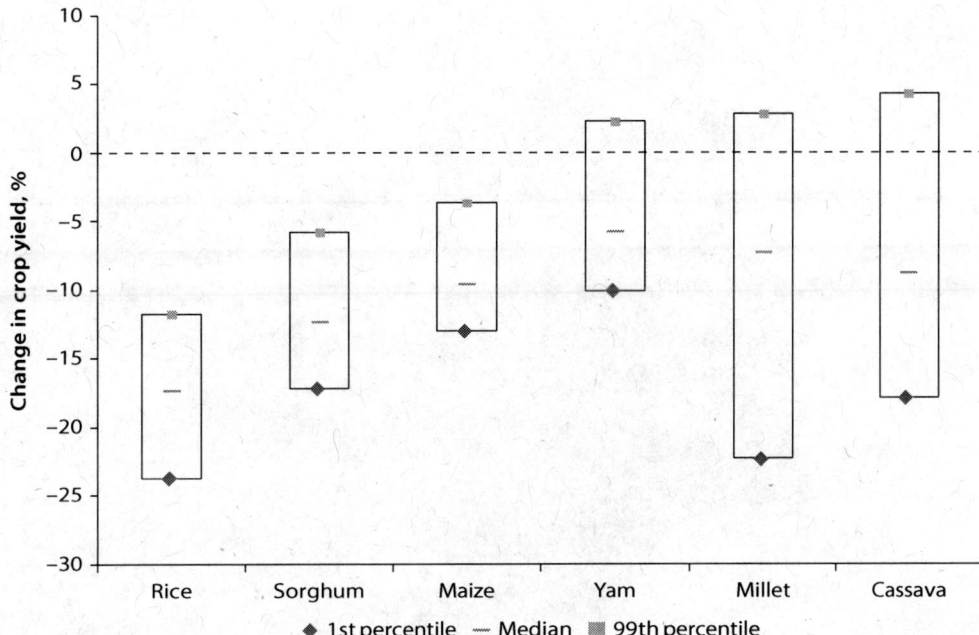

Source: Authors' calculations based on data sources listed in table 3A.1.
Note: The figure reports the range of yield change (compared to historical averages) produced by different climate models.

infrastructure will be required. The more significant impacts are expected in large swaths of the north and the southwest (map O.2).

Increased temperatures will affect livestock and rural livelihoods. The projected temperature increase is likely to trigger higher livestock mortality rates. Along with reduced yields from rain-fed crops, this is likely to have serious implications for livelihood and poverty in Nigeria. Map O.3 shows the distribution of risk for livestock sustainability (pasture) and suitability (thermal stress).

Sustainable Land Management Options

There is a wide range of sustainable land and water management practices that can offset, or even reverse, the effects of climate change on crops and livestock. These include conservation agriculture practices, such as integrated soil fertility management, water harvesting, minimal or no tillage, and agro-forestry; others comprise shifting planting dates, crop rotation, and restoration of degraded pastures. For seven practices that easily lent themselves to crop modeling, it was found that compared to a no-adaptation case, they could improve yields in 30–90 percent of the cases, depending on the crop, the time horizon, the climate model, and the agro-ecological zone considered.

Map O.2 Decline in Food Security by AESZ, 2050 Compared to 2000

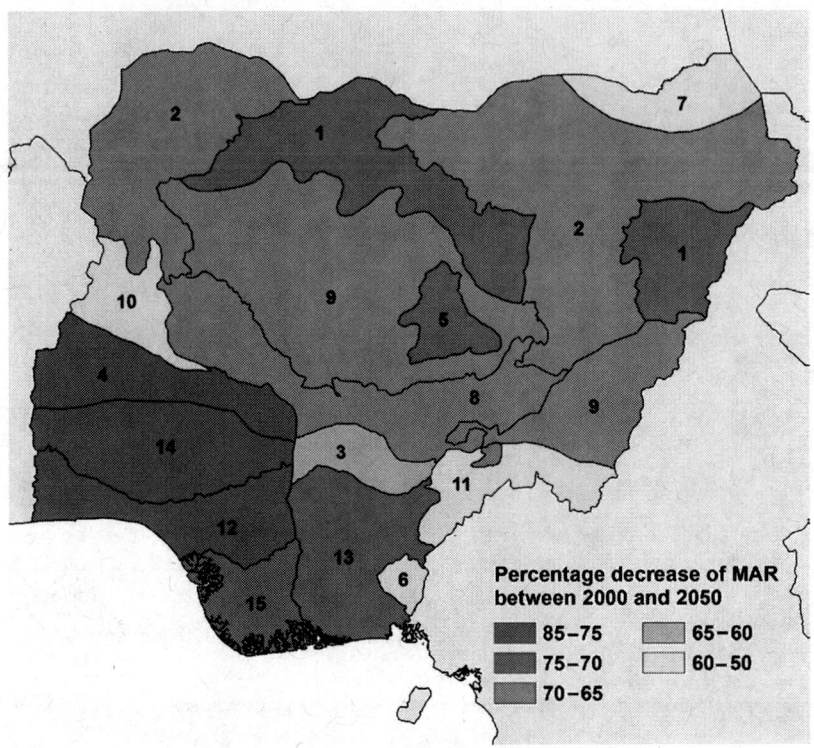

Source: Authors' calculations based on data sources listed in table 3A.1.
Note: AESZ = agro-ecological subzone. The figure maps the decrease in 2050 of the mean adequacy ratio (MAR), a standard
food security indicator.

Although nutrient management through manure and residues was found to be effective in more than half the cases, the choice of suitable adaptation options is still site- and crop-specific, with some areas (figure O.3) requiring a mix of up to four options, while for others one or two might be sufficient.

Since the effectiveness of adaptation solutions varies considerably in different climate scenarios, options should be identified that can perform well in as many scenarios as possible.

The Role of Irrigation

Robust adaptation in rain-fed areas can in many situations close the yield gap; if applied to 0.6–1 million hectares by 2020 it can eliminate most short-term climate impacts. In the longer-term, however, it is unlikely to eliminate climate effects entirely, particularly under scenarios of more severe climate impacts.

For these reasons, expansion of irrigation might be considered as a complementary strategy to enhance the resilience of agriculture—although achieving the government's current objectives, such as the 2.1 million ha covered by the

Map O.3 Integrated Risk for Livestock, 2050 Compared to 2000

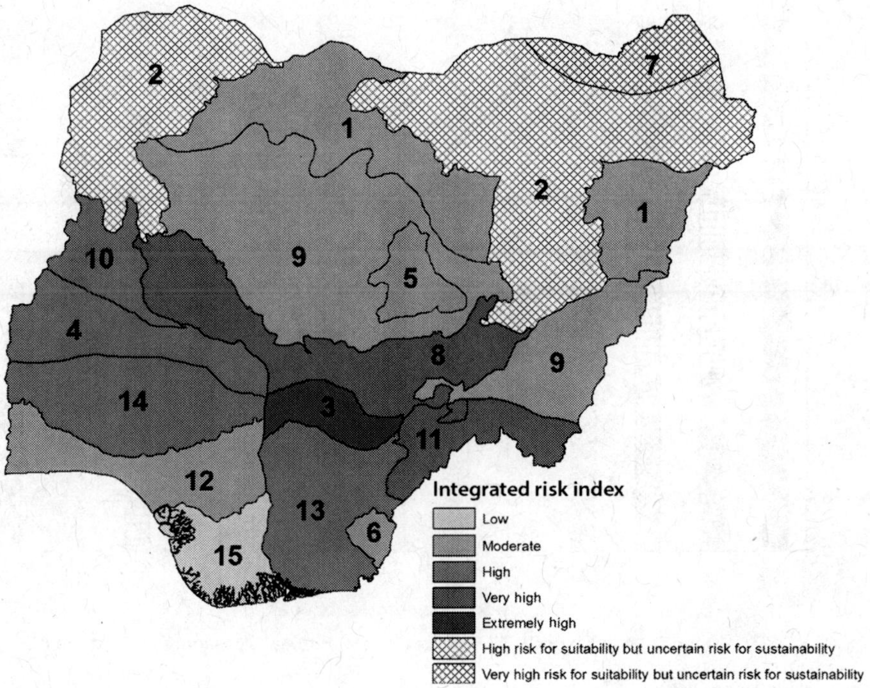

Source: Authors' calculations based on data sources listed in table 3A.1.
Note: Numbers refer to agro-ecological subzones.

Master Plan for Irrigation and Dam Development, will be important in itself to increase production and reduce the effects of climate change on food security. As a first approximation, it was found that by 2050 a combination of better management of 13–18 million rain-fed ha and expansion of irrigation to an additional 1.5–1.7 million ha can fully offset the output gap. If unit costs can be kept in check, such adaptation strategy is economically attractive, and an aggregate benefit-cost ratio ranging from 1.3 to over 3 can be achieved.

The Climate-Smart Way

This book explores the application (a first in Nigeria) of the robust decision-making approach to make irrigation development more resilient to climate change. Irrigation is vital to baseline sector development, but also (as mentioned earlier) to adaptation. However, irrigation entails large capital outlays, with unit costs often higher in Nigeria than in comparator countries. Adequate sizing of any given irrigation scheme (for example, the size of reservoirs for schemes based on dams) depends, among other factors, on the expected climate. In particular, for any given area to be irrigated, the expectation of drier conditions might require

Figure O.3 Robust Rain-Fed Adaptation Strategies

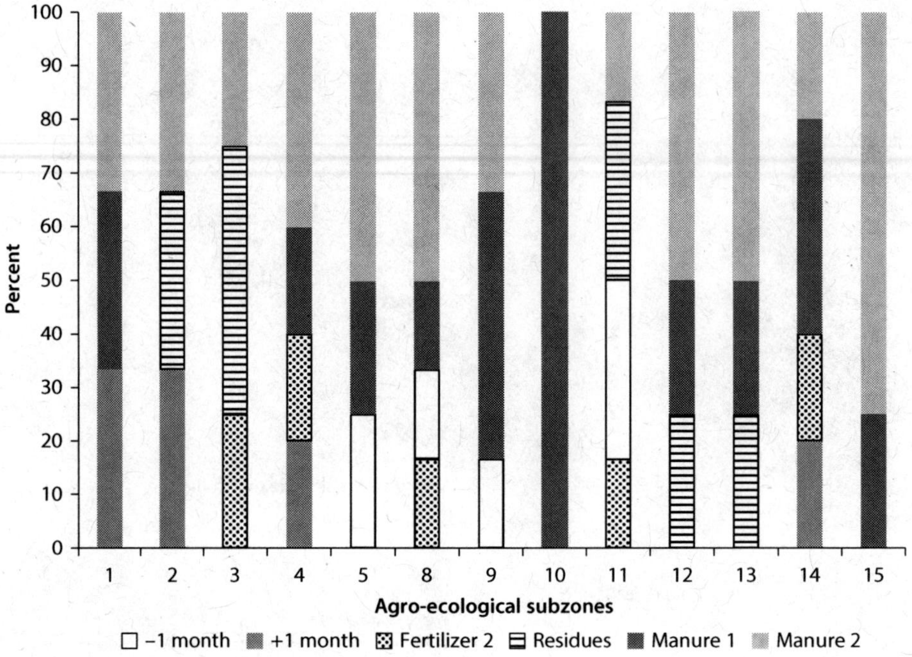

Legend: ☐ −1 month ▨ +1 month ▧ Fertilizer 2 ▤ Residues ■ Manure 1 ▨ Manure 2

X-axis: Agro-ecological subzones (1, 2, 3, 4, 5, 8, 9, 10, 11, 12, 13, 14, 15)
Y-axis: Percent (0–100)

Source: Authors' calculations based on data sources listed in table 3A.1.

more storage than the case in which a wetter climate is expected. This makes planning and design challenging, since it is not possible to know how climate will unfold in the future.

Using historical climate records as a basis for determining what investment is adequate in water storage or in area equipped for irrigation is likely to result in "regrets," because the investment will be undersized, if the climate turns out to be drier than expected; or oversized, if it is wetter. Analyzing 18 real-life irrigation projects, this assessment found that such regrets can be as large as 40 percent of investment costs. By selecting the investment strategy that minimizes the risk of misjudgments across multiple climate outcomes, regrets can be reduced by 30–50 percent—even 90 percent in some locations. The remaining regrets can be reduced by adding flexibility to the system: cropping patterns, water use, or other parameters can be adapted for wet or dry years to increase the return on irrigation investment. These results are illustrative because some site-specific information needed to strengthen the analysis was not readily available; for example, variables such as reservoir sedimentation, which already affects dams in Nigeria, could not be included.

Climate change must also be considered when planning new hydropower schemes. The overall feasibility of Nigeria's hydropower potential is not in question. On grounds of energy diversification and low carbon co-benefits, exploiting the entire 12 gigawatts (GW) of hydropower potential should be considered. However, given uncertainty about future precipitation and river runoff, it is not easy from an economic perspective to optimize hydropower schemes. Here, too, robust decision making could be considered. For example, a first-cut analysis of the planned Mambilla scheme indicates that the possibility of a drier climate means that there is a risk that the scheme will not deliver the intended amount of power. Designing the dam without consideration of climate change could expose the project to a regret (the cost of failing to deliver firm power) of up to 25 percent of capital costs; using a robust approach to design that increases storage in anticipation of a possibly drier climate reduces the possible regrets to 5 percent.

The Case for Acting Now

In Nigeria, as in many other countries, climate change impacts are likely to be significant, especially in the medium to long term. There are at least three reasons why the government may wish to act now to address them:

1. Many actions that will reinforce longer-term climate resilience will also help reduce vulnerability to current climate swings.
2. Investment decisions that will soon be made about long-lived, and expensive, infrastructure, such as irrigation or hydropower, will determine how resilient these investments will be to the harsher climate of the future.
3. Building the knowledge, capacity, institutions, and policies needed to deal with the climate of the future takes time.

The longer Nigeria delays action, the less time it will have to get ready, and rather than doing prevention it will have to find cures, which is typically more expensive and less effective.

Ten Ways to Enhance Climate Resilience by 2020

Nigeria has a number of actions and policy choices it might consider for building up its ability to achieve climate-resilient development. In the shorter term—action to be initiated by 2015 with targets reached by 2020—the Federal Government could give priority to 10 activities (table O.1) in the areas of institutions, information, and investments. Chapter 7 of the book expands on these suggestions and identifies complementary longer-term actions.

Institutions and Policies

1. The Federal Government's Economic Management Team could define priority adaptation actions, building on the 2011 National Adaptation Strategy and Plan of Action on Climate Change for Nigeria (NASPA-CCN) and the results of this book. This would ensure that enhancing climate resilience becomes a cross-cutting priority, not just a concern of the Ministry of the Environment; and that there are clear directions for coordinating, across institutional boundaries, the climate-related actions of different ministries, departments, and

Table O.1 Recommendations for Action by 2020

Recommendation	Lead agencies
Institutions and policies	
1. Define priority adaptation actions by sector (including on capacity building), to be endorsed by the Economic Management Team and integrated into federal programs.	Federal Ministries of Finance and Environment
2. Harmonize policies and legislation related to water resources management.	Federal Ministry of Water
Information and knowledge	
3. Launch a dedicated program of applied research on climate-smart agriculture (CSA).	Federal Ministry of Agriculture
4. Define an action plan for strengthening extension services through partnership and cost-sharing arrangements in five to ten states, including assistance to farmer organizations for accessing carbon markets.	Federal Ministry of Agriculture, state governments, commercial service providers, producer organizations, NGOs, and faith-based organizations
5. Prepare planning tools for climate-smart agriculture (e.g., a CSA atlas).	Federal Ministry of Agriculture
6. Define an action plan for strengthening the hydrometeorological system, with a 2020 target of increasing the current station density by 30–50 percent, and make data readily accessible to state governments.	Federal Ministry of Water, Nigeria Hydrological Services Agency
7. Prepare guidelines for designing climate-smart water infrastructure.	Federal Ministry of Water
Investments and resource mobilization	
8. Include in the government's Agricultural Transformation Agenda a program of CSA demonstration projects with a target of covering up to 1 million ha by 2020.	Federal Ministry of Agriculture
9. Pilot the use of robust decision making or similar techniques in the feasibility studies of specific irrigation and hydropower projects.	Federal Ministry of Water, Federal Ministry of Power
10. Put in place in a few states integrated watershed management and monitoring plans (accelerating current efforts, such as those supported by NEWMAP, JICA, and the EU).	Federal Ministry of Water and para-statals, Federal Ministry of Agriculture, Federal Ministry of Environment, state governments

agencies (MDAs). The priority adaptation actions should include significant efforts—to be sustained over time—to increase capacity, to ensure that climate resilience becomes part of the core competencies of relevant staff in MDAs.

2. The Federal Ministry of Water could lead efforts to consolidate and harmonize policies and legislation related to water resources management, perhaps by fast-tracking the parliamentary review of the Water Resources Bill, as a prerequisite for systematic and effective integration of climate change considerations into sector planning and development.

Information and Knowledge

3. The Federal Ministry of Agriculture could launch an applied research program on climate-smart agriculture (CSA), with individual grants for research to be awarded competitively to institutions in the National Agricultural Research System (NARS). The program could look at options such as improved seed varieties, changes in planting dates, low or no tillage, natural regeneration and agroforestry, pasturage management regimes such as rotational grazing, integrated soil fertility and nutrient management, and rainwater harvesting.

4. The Federal Ministry of Agriculture could draw up an action plan for strengthening extension services through partnership and cost-sharing arrangements in five to ten states, including assistance to farmer organizations for accessing carbon markets. The plan should be backed up by an agreement with the Ministry of Finance to provide federal resources ensuring the sustained functioning of extension over time.

5. The Federal Ministry of Agriculture could develop planning tools, such as a CSA atlas, to define and prioritize, across space and crops, opportunities for adopting "triple-win" agricultural options that have higher yields, higher climate resilience, and less emissions of greenhouse gases.

6. The Federal Ministry of Water and the Nigeria Hydrological Services Agency (NIHSA) could set out an action plan to enhance the hydrometeorological system, in terms of both the density of the observation network, and of the capacity at headquarters and in river basin agencies to organize and analyze data for decision making. A 2020 target might be to increase current station density by 30–50 percent. The data should be freely accessible by state governments.

7. The Federal Ministry of Water and NIHSA could publish guidelines for climate-resilient infrastructure for water storage and use, taking into account the full range of possible climate outcomes, so that hydropower and irrigation schemes are able to meet standards of service in as many climate scenarios as possible.

Investments and Resource Mobilization

8. The Federal Ministry of Agriculture could incorporate into the Agricultural Transformation Agenda (ATA) a program to promote triple-win climate-smart, sustainable land management practices in up to 1 million ha by 2020. This order of magnitude of adaptation efforts is necessary to offset short-term climate impacts on agriculture. The program could give priority to regions in the north and in the southwest that are particularly vulnerable; and to strategic crops and supply chains, such as rice, which appears vulnerable in many climate scenarios, and cassava, which at least in the medium term may be better suited to coping with a changing climate.

9. The Federal Ministries of Power and Water Resources could pilot robust decision making or similar techniques in feasibility studies for specific irrigation and hydropower projects, to ensure that their design is optimized to take into account a wide range of climate change possibilities.

10. The Federal Ministries of Water, Agriculture, and Environment, in collaboration with state governments, could put in place, in a few states, integrated watershed management and monitoring plans (accelerating efforts such as those supported by the World Bank under the Nigeria Erosion and Watershed Management Project [NEWMAP], the Japan International Cooperation Agency [JICA], and the European Union [EU]), to better integrate climate resilience into watershed management.

CHAPTER 1

Introduction

The Federal Government of Nigeria and the World Bank have agreed, as part of the Country Partnership Strategy (CPS) 2010–13, to undertake a comprehensive program of analytical work study to provide insights into what climate change implies for Nigeria's development agenda.

Challenges and opportunities related to low-carbon development are addressed in a separate volume (Cervigni, Rogers, and Henrion 2013). This book focuses on climate resilience. Building on the National Adaptation Strategy and Plan of Action on Climate Change for Nigeria (NASPA-CCN; BNRCC 2011), it evaluates the short- and medium-term risks (short-term up to 2020 and medium-term up to 2050) that climate change poses to Nigeria's agriculture and water resource management objectives as defined in Vision 20: 2020. These sectors have been chosen because they are currently strategic to the country's macroeconomic structure and are likely to remain of central importance to Nigeria's development in the foreseeable future.

The study's objectives, scope, and methodological approach were defined in a number of consultations held in 2010 and 2011 between the World Bank team and ministries, departments, and agencies (MDAs) of Nigeria's federal government. The objectives are to

- Assess climate-related risks based on available data and existing models, focusing on the two interacting sectors, agriculture and water resources, and investigate the related effects on higher-order policy variables, such as gross domestic product (GDP) growth, food security, and energy supply
- Identify and evaluate adaptation measures in the sectors analyzed
- Test innovative approaches to decision making under climate uncertainty, especially investment decisions about irrigation and hydropower.

This book considers the implications of climate change up to the middle of the current century; although climate conditions are expected to worsen later in the century, the period covered by this analysis makes it possible to address a sufficiently wide range of likely impacts and implications for development policies.

In this book, "agriculture" covers both crop cultivation and livestock, and "water resources" refers to the use of water for irrigation and hydropower. The analysis covers production of six food crops—sorghum, millet, maize, rice, cassava, and yams—the cereals and tubers that in 2008 represented 80 percent of agricultural value added (Nwafor *et al.* 2010).

The study makes no attempt to cover the full range of impacts that climate change may have on Nigeria's development plans. Since the analysis may also apply to other areas, such as human health, forests, coastal zones, fisheries, and water for domestic uses, the findings are likely to be an approximation that provides lower-bound estimates of the wider spectrum of effects that climate change may trigger.

The study did not analyze the Niger and Benue Rivers directly because most agricultural and water development is occurring or is planned along tributaries to these rivers rather than in the main stems themselves. The effects of climate change on the main rivers will depend heavily on what happens in countries upstream from Nigeria; they are extensively studied in a separate study under-way on the climate risks to the Sustainable Development Action Plan (SDAP) for the Niger basin (World Bank 2011).

The analysis is based on data and information collected up to June 2012; changes in government policies or other developments that have occurred since then are not reflected in the book.

The structure of the book is as follows. Chapter 2 provides essential back-ground on sectors of inquiry. Chapter 3 summarizes the methodology used in the analysis (with additional details to be found in the appendices). Chapter 4 discusses projections of future climate change and their uncertainty. Chapter 5 reports the findings of the analysis of climate change impacts. Adaptation options are analyzed in chapter 6. Chapter 7 presents conclusions and proposes policy recommendations.

References

BNRCC (Building Nigeria's Response to Climate Change). 2011. *National Adaptation Strategy and Plan of Action on Climate Change for Nigeria* (NASPA-CCN). Ibadan, Nigeria: BNRCC. http://nigeriaclimatechange.org/naspa.pdf.

Cervigni, R., J. A. Rogers, and M. Henrion, eds. 2013. *Nigeria: Opportunities for Low-Carbon Development.* Washington, DC: World Bank.

Nwafor, M., X. Diao, and V. Alpuerto. 2010. *A 2006 Social Accounting Matrix for Nigeria: Methodology and Results.* Nigeria Strategy Support Program (NSSP), Report NSSP007, International Food Policy Research Institute (IFPRI), Washington, DC.

World Bank. 2011. "Climate Change Brings Opportunity Alongside Challenges for Africa." Washington, DC. http://climatechange.worldbank.org/content/climate-change-brings-opportunity-alongside-challenges-africa.

CHAPTER 2

Background

The Federal Government of Nigeria (FGN) has put forward an ambitious vision for its economic development by 2020, known as Vision 20: 2020 (FGN 2010). The FGN considers it to be a platform for socioeconomic transformation that will in a few more years position Nigeria among the 20 largest economies in the world, with annual per capita income of not less than US$4,000. This will most likely require an acceleration of recent growth rates, which in the last decade have averaged 5 percent, although hitting close to 7 percent in 2009.

Vision 20: 2020 promotes a growth strategy that is built on, among other developments, (a) rapid expansion of the energy sector, particularly hydropower; (b) a sustained increase in agricultural productivity; and (c) diversification of the economy into manufacturing and services, which as yet are significantly underdeveloped.

As already recognized in Nigeria's First National Communication to the UN Climate Change Convention, and more recently in the National Adaptation Strategy and Plan of Action on Climate Change for Nigeria (NASPA-CCN) (box 2.1), current climate variability and future change have serious implications for the country's development prospects. They may well interfere with Nigeria's ability to achieve, and sustain over the longer term, the objectives of Vision 20: 2020. Climate shocks could prevent agriculture from reaching its productivity potential; change in rainfall patterns could affect generation of hydropower, which is likely to remain central to the country's energy mix. To better inform the analysis of impacts and identification of adaptation options, this chapter briefly reviews key aspects of Nigeria's agriculture sector and water resources.

Agriculture and Food Security

The agriculture sector, incorporating crop cultivation and livestock, is strategic for Nigeria's economy. It contributes more than 40 percent of gross domestic product (GDP) and accounts for about 70 percent of employment (NBS 2010).

Since 2000, agricultural growth has averaged 5.6 percent annually, but mainly from converting forests, bush land, wetlands, and woodlands into cropland.

Box 2.1 Documentation of Nigeria's Vulnerability to Climate Change

Nigeria's First National Communication submitted a report in 2003 to the United Nations Framework Convention on Climate Change (UNFCCC) that outlined its main vulnerabilities:

- The heavy dependence of the economy and food security on rain-fed agriculture makes Nigeria highly susceptible to fluctuations in rainfall and rises in temperature.
- The energy sectors are sensitive to climate variability and change; in particular, Nigeria's hydropower potential is deeply affected by variations in rainfall.
- Rapid population growth (almost 3 percent annually) is coupled with pervasive poverty, which reduces resilience to multiple climate risks.

The FGN and a number of civil society organizations in 2011 formulated the National Adaptation Strategy and Plan of Action on Climate Change for Nigeria (NASPA-CCN; BNRCC 2011), with the intent of reducing Nigeria's vulnerability to the negative impacts of climate change by improving local and national adaptive capacity and resilience, leveraging new opportunities, and facilitating collaboration with the global community.

Based on a review of published climate projections—which point to increasing warming but more uncertain rainfall trends although with higher variability in precipitation—expected impacts were reviewed for a wide range of themes in strategic sectors: agriculture (crops and livestock), water resources (including coasts and fisheries), forests, biodiversity, health and sanitation, human settlements and housing, energy, transportation and communication, disaster mitigation and security, livelihood, vulnerable groups, and education.

For each, four main climate change–related hazards are considered: (a) higher temperatures; (b) change in the amount, intensity, and pattern of rainfall; (c) extreme weather events, including sea surge and drought; and (d) a rise in the sea level.

As an example, the adverse impacts of climate change are expected to both lead to production losses in agriculture and affect the characteristics of the freshwater resources on which Nigerians depend. The impacts will vary depending on the agro-ecological zone (AEZ), production, and the sociocultural conditions for any given area of Nigeria.

To guide in reducing the impacts of climate change through adaptations that can be undertaken by federal, state, and local governments; civil society; the private sector; and communities and individuals, the NASPA-CCN outlines recommended strategies for each priority sector and themes, and defines related policies, programs, and measures.

Meanwhile, yields of the main crops have been flat for the past two decades. Because of conversions and other factors, land degradation[1] persists in all major agro-ecological systems, constraining yields.

The main components (box 2.2) in Nigeria's food basket are cereals and tubers, such as rice, maize, corn, millet, sorghum, yams, and cassava. Together they account for 80 percent of the sector's value added.

About 44.5 percent of land area was being cultivated in 2009;[2] of the land under cultivation, 41.2 percent consisted of arable lands and 3.3 percent of permanent crops.[3] About two-thirds of the cropped areas are located in

Box 2.2 Nigeria's Primary Crops

Crops in Nigeria are divided between food crops and export crops. The most important food crops are yams and cassava in the south and sorghum (Guinea corn) and millet in the north. Cocoa beans dominate exports at about 65 percent of trade, followed by groundnut oil, palm nuts and oil, and kernel corn.

Sorghum is the most important cereal food in Nigeria, in particular in the north. At more than 7.6 million hectares (ha) under cultivation in 2008, it occupies over 40 percent of the total area devoted to cereals (FAOSTAT). Nigeria is the second largest sorghum producer in the world.

Millet is the second most important cereal in Nigeria, with a cultivated area of about 5 million ha and total production of more than 9 million tons in 2008 (FAOSTAT). It is grown and eaten in particular in the northern savanna.

Nigeria is the also the tenth largest producer of maize worldwide, and the main producer in tropical Africa, with annual production of more than 6 million tons. As reported by USAID MARKETS (2010), maize will continue to play a large and important role in Nigeria's food production, with 3.8 million ha cultivated. This crop has several advantages over other crops: it is a major source of energy; it is usually the first crop to be harvested for food during the hunger period; it is easy to grow, whether alone or intercropped; and it is easy to harvest. Industrial demand for it is also increasing, particularly in the food, beverage, and livestock feed industries.

Rice is the other important cereal cultivated in Nigeria, with 2.4 million ha of harvested land. Rice production has emerged as the fastest-growing subsector and most sought-after commodity in the Nigerian food basket. Because of supply and demand gaps, imports have soared to an unprecedented volume of nearly 1 million metric tons a year, costing about US$1 billion. Rice is cultivated in virtually all of Nigeria's AEZs, from the mangrove and swamp ecologies of the Niger delta in the coastal areas to the dry zones of the Sahel in the north.

Rice yields are low, averaging 1.7 tons per hectare. Rain-fed lowland rice is the predominant production system, accounting for nearly 50 percent of the total rice-growing area in Nigeria; 30 percent of production is rain-fed upland rice, while just 16 percent is high-yielding irrigated rice. Other production systems make up the remaining 4 percent (USAID MARKETS 2009).

Cassava and yams are the lead crops for the Nigerian economy. Nigeria produces more of both than any other country in the world. Cassava is abundant in 24 of the 36 states, requires minimum labor and inputs, and is at the center of food security for millions of Nigerians. As reported by USAID MARKETS, Nigeria produces more than 45 million tons of cassava a year. Yet the full yield potential has not been realized because smallholder production rarely exceeds 11 tons per ha. Compared to the cassava yields of Malawi (22 tons per ha) or India (36 tons), Nigeria has a clear opportunity for growth. The introduction of improved varieties and the adoption of best agronomic management practices can make it possible to achieve even 50 metric tons (MT) per ha. Farmers working with the USAID-funded Cassava Enterprise Development Project implemented by the International Institute of Tropical Agriculture achieved average on-farm yields of cassava of about 25 MT per ha (http://www.nigeriamarkets .org/files/Cassava%20fact%20sheet_FINAL.pdf).

box continues next page

Box 2.2 Nigeria's Primary Crops *(continued)*

The crop is efficient in producing carbohydrates; tolerant to drought and impoverished soils, even though it thrives best on fertile, sandy-clay soils; and is very flexible in terms of planting and harvesting times. For these reasons, cassava is especially essential for food security in regions prone to drought and with poor soil. It is the world's fourth most important staple after rice, wheat, and maize and is important to the diets of over 1 billion people (from http://www.fao.org/ag/agp/agpc/gcds/). Cassava produces best when rainfall is fairly abundant, but it can be grown where annual rainfall is as low as 500 mm or as high as 5,000 mm. Its ability to withstand prolonged periods of drought makes it valuable in regions where annual rainfall is low or seasonal distribution is irregular.

The Food and Agriculture Organization of the United Nations (FAO) reported that in 2007 (FAOSTAT), worldwide yam production amounted to more than 47 million tons, of which Africa produced 96 percent. Most of the world's production—92 percent—comes from West Africa, with Nigeria alone producing 66 percent, equaling more than 31 million tons. African countries imported almost 7,500 tons in 2007, and exported almost 22,400, of which Nigeria exported less than 1 percent.

Yams are a high-value food crop that is easily grown and matures quickly in the right soil conditions. Unlike most other tropical root crops, it has good keeping qualities and may be harvested well in advance of eating. Most varieties grow best in areas with rainfall higher than 1,500 mm/year and require at least a 6-month growing season with well-distributed rainfall.

---

the north, with the rest about equally distributed between the center (Middle Belt) and the south. With irrigation accounting for a negligible fraction of cultivated area, rainfall has a heavy influence on national crop production. Cultivation calendars and cropping patterns are different in the north compared with the south, largely because of the differences in precipitation.

Recent climate patterns (see, e.g., Nigerian Meteorological Agency's [NIMET] 100-year database or Lebel and Ali 2009) adversely affected national crop production. Increasingly severe crop failure or loss of yields due to the false start of the rains, frequent dry spells during the growing seasons, and early cessation of rains limit the growing season (Adejuwon 2008; Odekunle 2004); crop damage from storms and flooding, rising temperatures, and pest infestations also undermine crop production. Crop failures and yield losses thus jeopardize nutritional status, and public health.

Smallholders control 80–90 percent of Nigeria's farms (map 2.1). Smallholder farmers live in areas with limited access to pesticides, fertilizers, hybrid seeds, irrigation, and other productive resources. Consequently, farming is inefficient, and there is a regular shortfall in national domestic production. Food imports account for some 10 percent of total national imports (NBS 2010).

The diet of many Nigerians does not meet basic nutritional requirements. The average daily intake of 9 g of protein (Oluleye and Osunfuyi 1991) is far below the recommended rate of 40 g (WHO 1985) and grossly inadequate to sustain

Map 2.1 Nigeria: Spatial Distribution of Farms by Classes, 2007

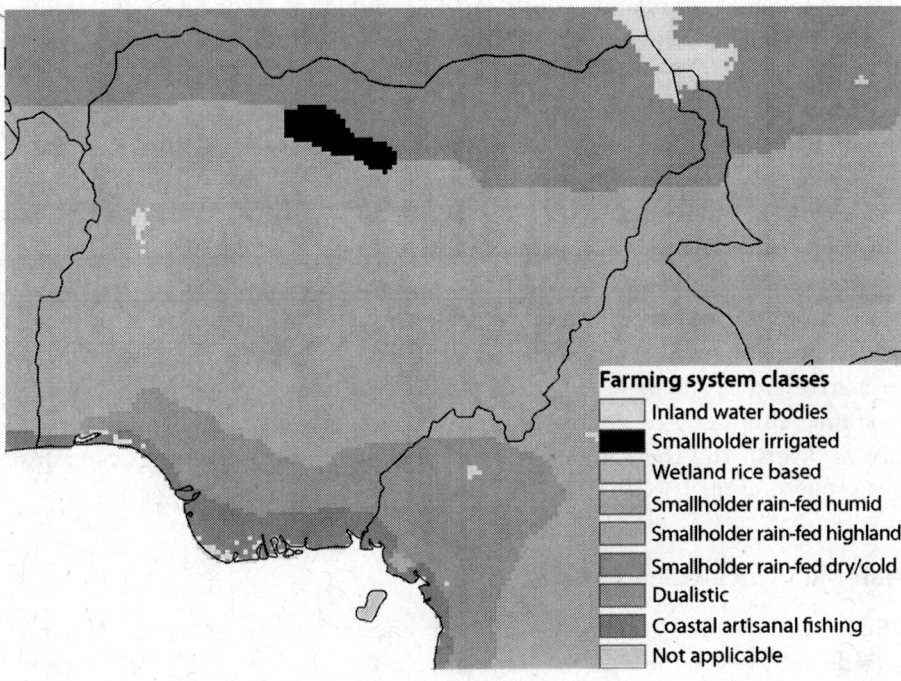

Farming system classes
- Inland water bodies
- Smallholder irrigated
- Wetland rice based
- Smallholder rain-fed humid
- Smallholder rain-fed highland
- Smallholder rain-fed dry/cold
- Dualistic
- Coastal artisanal fishing
- Not applicable

Source: Authors' calculations based on data sources listed in table 3A.1.

good health. It is not a surprise that the most vulnerable members of Nigerian society suffer from a diversity of debilitating illnesses.

Nigeria is not only listed by FAO (AQUASTAT; FAO 2005) among the nations that are technically unable to meet their food needs from rain-fed production at a low level of inputs, but it is also likely to remain so even at intermediate levels of inputs for 2000–25.

Government policies attach high priority to increasing agriculture productivity in order to help reduce poverty and achieve objectives for food security and diversification of the economy away from oil (Ephraim *et al.* 2010; NPC 2004). As outlined in the Vision 20: 2020 strategy and other sector strategy documents, government goals include:

- Triple domestic agricultural productivity by 2015 and double that again by 2020 to make Nigeria self-sufficient in food and fiber requirements
- Reduce current food imports by 50 percent by 2013
- Derive over 50 percent of the nation's foreign exchange earnings from agro-industrial exports by 2013.

Livestock

Livestock management contributes about 5 percent of total Nigerian GDP (Bénard, Bonnet, and Guibert 2010). Nigeria is one of the four leading livestock producers in Sub-Saharan Africa. The most recent statistics (NBS 2010) report that the livestock population consists of 16.5 million head of cattle, 56.5 million goats, 35.5 million sheep, and 192 million poultry. Data from FAO for 2006 (FAOSTAT) show that while beef and veal, goat, and game meat production increased only gradually between 1995 and 2004, production of sheep meat doubled. Large numbers of live cattle, sheep, and particularly goats are imported.

Livestock is vulnerable to climate change and variability. Animals are heavy consumers of water and fodder, the availability of which is declining. Thus, warming can negatively impact animal health, productivity, and water requirements. NASPA-CCN (BNRCC 2011) reports how a shorter rainy season is affecting the amount of water available for cattle and the quality of grazing fields in some pastoral communities in northwestern Nigeria. These communities are particularly vulnerable to loss of livestock because they currently have few economic alternatives. They are already practicing such local adaptation strategies as diversifying the composition of their herds and harvesting water from their zinc rooftops (BNRCC 2011).

Water Resource Management

With about 1,800 m³/capita/year of total renewable water resources, Nigeria is considerably below the Sub-Saharan Africa average of about 6,500, although it is well above the 1,000 threshold typically used to define water scarcity. Nigeria must manage these relatively limited water resources effectively if it is to reach its development objectives. Yet relatively few resources are presently being directed to critical development priorities, such as hydropower or irrigation.

Nigeria's total annual renewable water resources are estimated (FAO 2005) at 286.2 km³, 77.2 percent of which is produced internally and the rest is surface water coming from neighboring countries. Exploitable surface water resources are estimated at about 96 km³/year. Annual accessible groundwater resources are estimated to be 60 km³, distributed 17 percent in the north, 43 percent in the middle, and 40 percent in the south. Dam capacity nationwide is estimated at 45.6 km³. Water from other sources, such as wastewater (reused, treated, produced), desalination, and reused agricultural drainage water is not significant.

Nigeria has 106 large dams and 120 medium and small ones. Most of them serve multiple purposes (water supply, irrigation, and hydropower). In terms of water withdrawal, estimated at 8 km³/year (5 percent of total exploitable water resources), 69 percent goes for agriculture, 21 percent to households, and 10 percent to industry.

To support water resources management, the Japan International Cooperation Agency (JICA 1995) produced a National Water Resources Master Plan (NWRMP) to help assure optimal water use and provide the short-term (to 2000) and longer-term (to 2020) development scenarios for meeting predicted regional social and economic demand. According to the NWRMP projections, to meet water demand between 2012 and 2020, incremental water storage of 2 billion cubic meters a year would be required.

JICA is currently supporting the Federal Ministry of Water Resources as it reformulates the master plan. A catchment management plan is also being drafted, initially for two priority hydrological areas and then for the remaining six.

Irrigation

Given the limited size of effectively irrigated areas (less than 1 percent of the cultivated area), the contribution of irrigated agriculture to total crop production is at present almost negligible (0.9 percent of total national production of grains and 2.3 percent of production of vegetables). However, the World Bank (2010) considers Nigeria one of the African countries with the most potential for expanding irrigation, which is likely to be vital if the sector is to reach the government growth targets.

According to the report of the International Commission on Irrigation and Drainage (ICID), Nigeria has three main categories of irrigation development (ICID 2011): (1) schemes under government control (formal irrigation); (2) farmer-owned and operated schemes (informal irrigation) that the government supports with subsidies and training; and (3) residual floodplains, where no government aid is supplied and irrigation is based on traditional practices.

The area currently equipped for irrigation is 293,117 ha, but area actually irrigated in large and medium-scale schemes adds up to only 119,350 ha. The traditional Fadama-type irrigation accounts for 181,000 ha, which brings the total to 300,350 ha (Enplan Group 2004). The most important irrigated crops are rice, wheat, and vegetables, which together occupy 90–95 percent of the total water-managed area, irrigated plus flooded.

Since precipitation and cropping patterns differ considerably across agro-ecological areas, and potential to improve yields by irrigation is highly variable, to ensure effective management of water resources Nigeria must achieve a strategic balance between rain-fed and irrigated production. Rain-fed production can be stabilized at low unit cost but will always be vulnerable to drought. Irrigated production can buffer the impacts of drought where it can draw on stored surface water, groundwater, or both.

The costs of not taking advantage of irrigated production are high. Surging food import bills are draining foreign exchange. The National Economic Empowerment and Development Strategy (NEEDS) anticipates an agricultural development program to drastically reduce food imports and boost agricultural exports through stabilization and expansion of rain-fed production, intensified through irrigation, and accelerated commercialization with private sector help.

A number of government policies (e.g., the Master Plan for Irrigation and Dam Development and the National Fadama Development Project [NFDP]) encourage a viable structure of public and private irrigation with a balance of small-, medium-, and large-scale irrigated production. Rehabilitation and extension of public schemes are priorities. The master plan also proposes construction of new dams and irrigation schemes to improve irrigation infrastructure. Parallel activities include capacity building and service provision, improvement in institutions and rural infrastructure, and facilitation of private sector engagement.

Hydropower

Among the problems of Nigeria's power sector are inadequate access to the grid (estimated at only 40 percent); insufficient generation capacity to meet demand; shortages of gas for generating power; and an inefficient transmission and distribution system, which together add up to unreliability and frequent load-shedding. For reasons like these, an estimated 50–70 percent of electrical energy is currently produced off-grid by diesel and gasoline generators.

The current installed capacity of grid electricity is about 6,000 megawatts (MW), of which 65–67 percent is thermal and the rest water-based. Until 1960, power production in Nigeria was mainly from coal. Construction of the first hydropower station in Nigeria began in 1964 at Kainji, along the river Niger. The Kainji plant has an installed capacity of 760 MW. The tail water from Kainji Dam was then used to generate 540 MW at Jebba Dam, 97 km downstream. The third hydropower station, the Shiroro Dam, was commissioned in 1990 and has an installed capacity of 600 MW, bringing Nigeria's total capacity to 1,900 MW. Between 1990 and 1999, no new power plant was built, and the government seriously underfunded both capital projects and routine maintenance.

While trends (figure 2.1) confirm that power generation increased annually from 1971 to 2004 (about 97 percent supplied by public companies), installed capacity is still low, and the demand-supply gap is widening. Poor energy supply has forced many industrial customers to install their own generators, which have high costs to both companies and the Nigerian economy as a whole.

The situation is compounded by the failure of both hydro and thermal power stations to operate at full capacity. Zarma (2006) and Jimoh (2010)

Figure 2.1 Annual Hydroelectric Production in kWh and as a Share of Total Energy, 1971–2004

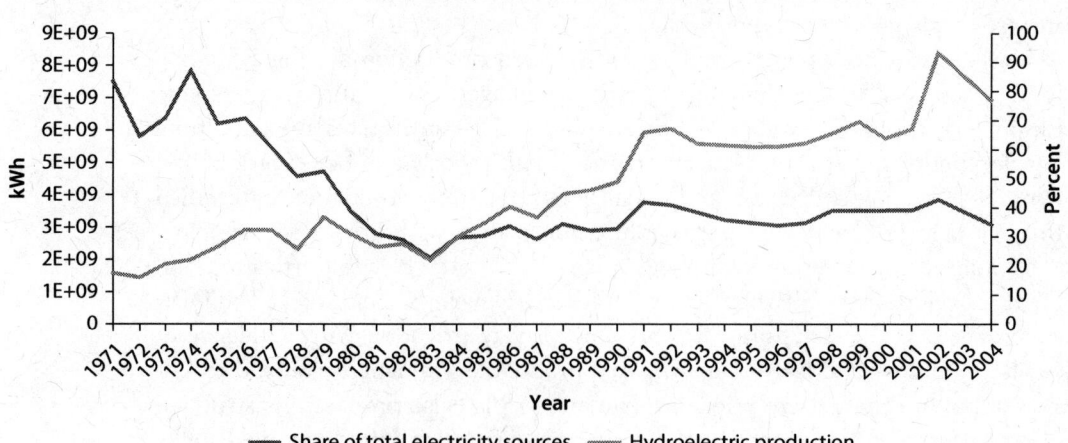

Source: Authors' calculations based on data sources listed in table 3A.1.
Note: kWh = kilowatt hour.

attribute the inability of the hydropower stations to operate at their installed capacity to

- Hydrological factors such as (a) seasonal variations in flow to the reservoir; (b) inter-year variations in flow to the reservoir; (c) conflict between competing users; (d) inefficient operations policies; and (e) reservoir sedimentation; and
- Nonhydrological factors such as (a) lack of maintenance and spare parts; (b) inadequate funding; (c) lack of qualified personnel; and (d) inadequate sector policies.

Since Nigeria has developed only 23 percent of its feasible hydropower (Manohar and Adeyanju 2009), which is an estimated 12,220 MW (from both small and large-scale plants), tapping the unused hydropower potential is crucial to achieve the sector's longer-term development objectives. Integrating climate resilience considerations into hydropower investment decisions therefore has strategic development significance.

The Lagos Metropolitan Area

The Lagos metropolitan area is essential to the Nigerian economy: Lagos accounts for 10 percent of the population but 20 percent of national GDP. Although this book does not address coastal zones and urban development, it is essential to recognize Lagos as an area that may be highly vulnerable to climate change and in need of follow-up work.

Much of the territory of Lagos city is spread over barrier islands or in close proximity to the seafront and the edge of the Lagos lagoon, which is connected to the sea by a number of channels. The northern part of the city is on the Ogun River floodplains. In 2006 the population was estimated to be 17.5 million. Based on the median UN population growth projection, by 2050 there are likely to be 46 million people living in Lagos. The current built-up area covers some 1,000 km² located in both Lagos and Ogun states. A large proportion of the population lives in slum areas, mainly on the lower flood-prone areas and some even on floating slums in the lagoons.

Climate variability affects Lagos city and its economic activities in numerous ways. Rising sea levels erode the coast and create flooding. Awosika and colleagues (1992) estimated that a rise in the sea level of 0.3 meter would affect 1–2 million people. More intense rainfall and river flows in the Ogun River, exacerbated by insufficient drainage due to rapid and unplanned growth of built-up areas and slums, will likely increase flooding that damages or destroys property and infrastructure, disrupting economic activities.

Since the construction over 100 years ago of structures to protect the Lagos harbor, natural erosion of Bar Beach on Victoria Island, which contains some of Nigeria's most expensive real estate, has intensified. It is estimated that coastal erosion already destroys 25–30 meters every year; projected rises in sea level and storm surges will accelerate the process. If no adaptation measures are put in place, up to 1,100 km² of land could be lost in future decades due to erosion along the highly desirable Lagos seafront (Awosika *et al.* 1992).

Recent climate change scenarios for Lagos project more intense rainfall. A large part of Lagos already suffers from flooding during normal rainfall, mainly because of insufficient capacity and maintenance of sewer and storm drains, and because many settlements are built on wetlands. Lagos will need significant improvements in drainage to support a larger population in a changing climate.

Notes

1. Given Nigeria's different agro-ecologies, land degradation is experienced mainly as soil erosion and infertility, desertification, and loss of forest and other vegetation.

2. FAOSTAT at http://www.fao.org/countries/55528/en/nga/.

3. This excludes lands for both livestock and wood and timber production.

References

Adejuwon, J. D. 2008. "Vulnerability in Nigeria: A National-Level Assessment." In *Climate Change and Vulnerability*, edited by Neil Leary, Cecilia Conde, Jyoti Kulkarni, Anthony Nyong, Juan Pulhin, 198–217. London: Earthscan.

Awosika, L. F., G. T. French, R. J. Nicholls, and C. E. Ibe. 1992. "The Impact of Sea Level Rise on the Coastline of Nigeria." Proceedings from the Intergovernmental Panel on Climate Change Symposium, "The Challenges of the Sea," Margarita, Venezuela, March 9–13.

Bénard, C., B. Bonnet, and B. Guibert. 2010. "Demand for Farm Animal Products in Nigeria: An Opportunity for Sahel Countries?" *Grain de Sel* 51, Special Issue, Nigeria: 14–15. http://www.inter-reseaux.org/revue-grain-de-sel/51-special-issue-nigeria /article/demand-for-farm-animal-products-in.

BNRCC (Building Nigeria's Response to Climate Change). 2011. *National Adaptation Strategy and Plan of Action on Climate Change for Nigeria* (NASPA-CCN). Ibadan, Nigeria: BNRCC. http://nigeriaclimatechange.org/naspa.pdf.

Enplan Group. 2004. "Review of Public Sector Irrigation in Nigeria." Federal Ministry of Water Resources/UN Food and Agriculture Organization. ftp://ftp.fao.org/AGL /AGLW/ROPISIN/ROPISINreport.pdf.

Ephraim, N., J. Pender, E. Kato, O. Omobowale, D. Phillip, S. Ehui. 2010. "Options for Enhancing Agricultural Productivity in Nigeria." Background Paper No. NSSP 011, January. Available at http://www.ifpri.org/sites/default/files/publications/nsspbp11 .pdf.

FAO (Food and Agriculture Organization). 2005. *Irrigation in Africa in Figures: AQUASTAT Survey 2005.* FAO Water Report 29 (with CD-ROM), Rome.

FGN (Federal Government of Nigeria). 2010. *Nigeria Vision 20: 2020: The First NV20: 2020 Medium-Term Implementation Plan (2010–2013).* Vol. 1. *The Vision and Development Priorities.* Lagos, Nigeria.

ICID (International Commission on Irrigation and Drainage). 2011. "Nigeria Country Profile." http://www.icid.org/cp_nigeria.html.

JICA (Japan International Cooperation Agency). 1995. "The Study on the National Water Master Plan" (Sector Report Vol. 2). Report prepared for the Federal Ministry of Water Resources and Rural Development, Nigeria.

Jimoh, O. D. 2010. "Operation of Hydropower Systems in Nigeria." Paper presented at a Seminar at the University of Ilorin, Ilorin, Nigeria, August 10.

Lebel, T., and A. Ali. 2009. "Recent Trends in the Central and Western Sahel Rainfall Regime (1990–2007)." *Journal of Hydrology* 375: 52–64.

Manohar, K., and A. A. Adeyanju. 2009. "Hydro Power Energy Resources in Nigeria." *Journal of Engineering and Applied Sciences* 4: 68–73.

NBS (Nigeria National Bureau of Statistics). 2010. "The Review of the Nigerian Economy." http://www.nigerianstat.gov.ng/.

NPC (National Planning Commission). 2004. *Meeting Everyone's Needs—National Economic Empowerment and Development Strategy.* Abuja, Nigeria: NPC.

Odekunle, T. O. 2004. "Rainfall and the Length of the Growing Season in Nigeria." *International Journal of Climatology* 24: 467–79.

Oluleye, O. B., and E. O. Osunfuyi. 1991. "Improving Productivity in the Livestock Sector in Nigeria." Lagos, Nigeria: Macmillan.

USAID MARKETS. 2009. "Package of Practices for Rice Production." http://www.nigeriamarkets.org/.

———. 2010. "Package of Practices for Maize Production." http://www.nigeriamarkets.org/.

WHO (World Health Organization). 1985. *Energy and Protein Requirements: Report of a Joint FAO/WHO/UNU Expert Consultation.* WHO Technical Report Series 724, Geneva, Switzerland.

World Bank. 2010. *Africa Infrastructure: A Time for Transformation.* Washington, DC: World Bank.

Zarma, I. H. 2006. "Hydro Power Resources in Nigeria." A country position paper presented at the 2nd "Hydro Power for Today" conference, International Centre on Small Hydro Power (IC-SHP), Hangzhou, China, April 22–25.

CHAPTER 3

Methodology of Analysis

This book builds on an analysis by the National Adaptation Strategy and Plan of Action on Climate Change for Nigeria (NASPA-CCN; see box 2.1) to add insights on climate impacts and possible responses in agriculture (crop and livestock) and water resources (irrigation and hydropower). The approach was designed to

- Establish a reference development scenario—the basis for assessing climate change impacts—that assumes no climate change and that incorporates sector objectives consistent with current government policies
- Define a range of possible future climate outcomes to reflect the disagreement among current climate models
- Evaluate climate impacts at multiple spatial scales by adopting different units of analysis (e.g., agro-ecological zones [AEZs], agro-ecological subzones [AESZs], hydrological areas [HAs], etc.; see appendix A), depending on the type of impact being investigated
- Pilot an innovative approach (robust decision making) to evaluate choices under deep climate uncertainty. This approach has been applied to adaptation decisions in irrigation and to a lesser extent hydropower.

Subject to the caveats discussed in box 3.1, the study team dedicated special efforts to collecting the data best suited for the analysis in terms of time and space resolution, from a combination of the international and national sources listed in annex 3A.

Climate Projections and Uncertainty

Past climate impact assessments have been plagued by inadequate treatment of uncertainty (box 3.2). To avoid this shortcoming and better assess the range of future climate variability, extremes, and ultimate impacts, a high-resolution regional climate model (RCM) was used to simulate and project climate changes from 1971 through 2065 under an A1B emission scenario, which represents a median between the most extreme (optimistic and pessimistic) story-lines developed in the Intergovernmental Panel on Climate Change (IPCC)

Box 3.1 Sources and Quality of Data Used in the Analysis

The majority of data used for modeling in this study were obtained from global datasets at high resolution (e.g., on the order of 1 km for soil and elevation; see annex 3A). Because the organizations supplying the data typically apply quality control protocols, the data were judged suitable for nationwide modeling.

Some of the data used for calibration and validation of models come from national observation networks and experimental studies. As in many developing countries, the degree of accuracy and representativeness of Nigerian data is constrained by the limited financial, institutional, and human resources available to collect, control, and organize such data. The findings of the analyses presented in this book are likely to be affected by those limitations.

The problems can be illustrated with regard to two types of data, on crop yield and runoff. Yield data, which are available at the level of states, tend to be too coarse to represent finer-scale interactions between climate, soil, and other factors. The study therefore had to complement this type of information with plot scale measurements in locations that may not necessarily be representative of the whole Nigerian territory. Such data were derived mainly from a literature survey; a comprehensive national database of long-term series (at least 10–15 years) would have been useful.

Obtaining adequate data for the rainfall-runoff model proved challenging because there were different time frames for precipitation data (1975–2009) and runoff station data (1960–89); both datasets had large temporal gaps (inter- and intra-annual), and the location and spacing of the monitoring gauges were uncertain. There were virtually no metadata on data quality or information about data reliability.

There is a real need to improve data collection networks and devote resources to assembling and digitizing current data from complementary national sources (projects, surveys, etc.) using shared protocols and formats. Improving data quality is essential for better climate change impact analysis.

Box 3.2 Climate Model Uncertainty

The synthesis chapter (Smith *et al.* 2001) of the Third IPCC Assessment Report emphasized the limitations of traditional impact assessments when a few climate models were used to evaluate how a system would respond to future changes in climate. Because it involves different regions, sectors, and resources, vulnerability assessment is complex in itself. Moreover, uncertainty increases in the process of evaluating risk. However, applying the concept of *ensemble forecasting* (e.g., Araújo and New 2007), it is possible to handle uncertainties associated with climate risk analysis (CRA) by using a range of scenarios for a given impact rather than relying on just one. While no individual model in the ensemble can be viewed as a more or less likely representation of the future, taken as a whole, the simulation ensemble helps define the *range* of possible changes.

Fourth Assessment Report (AR4). The selection of a single emission scenario is motivated by the common observation in the literature (Lionello *et al.* 2012; Olesen *et al.* 2007) that in the medium term uncertainty stemming from the choice of climate models is larger than the uncertainty associated with emission scenarios, which become more important in the long term; that is, beyond the time horizon of the present study.

The CMCC-MED global model output (about 80 km of horizontal resolution; Scoccimarro *et al.* 2011) was used for boundary conditions to run an RCM, COSMO Model in Climate Mode (COSMO-CLM; about 8 km of horizontal resolution; Rockel, Will, and Hense 2008). After validation with observed climate over the historical period, the RCM was bias-corrected for the whole simulated time frame (appendix B). Indeed, it is well known that outputs from climate models cannot be used to force impact simulations without some form of preprocessing to remove existing biases (Haerter *et al.* 2011). To remove biases, monthly scaling factors were calculated from the difference (temperature) or ratio (precipitation) in 30-year average monthly means between the observation dataset and regional simulation outputs. In principle, bias correction methodologies act on model output so that the main statistical properties of the corrected data match those of the observations.

Impact assessments depend on the climate scenarios available at temporal and spatial scales of relevance to the regional issues of importance. These scales are commonly far finer than the resolution of General Circulation Models (GCMs). Consequently, there is growing demand for regional-scale scenarios, often based on RCMs. Performing multiple high-resolution RCM simulations after changing boundary conditions as dictated by several GCMs may provide valuable insights into future climate uncertainty, but the method is very demanding in terms of both time and computational infrastructure. The approach chosen, which was to downscale projections from available GCMs, consisted of obtaining future climate anomalies (difference in monthly climate between RCM and GCM) from multiple GCMs and applying them to daily "reference" regional data from a single RCM simulation.

In practice, to capture the range of possible future climate outcomes, maintain high resolution, and take into account uncertainty about future climate, multiple climate projections from GCMs were used to "perturb" bias-corrected RCM results for 2006–65. Further, the projections were perturbed with spatial and monthly variations from the GCM (appendix C). For temperature perturbation, a monthly anomaly based on the difference between the GCM and the RCM was added to daily values; for precipitation, the scaling factor applied to the daily amount was the ratio between the monthly GCM and RCM values (Buishand and Lenderink 2004).

Nine global GCM simulations, part of the well-developed Coupled Model Intercomparison Project (CMIP3) experiment, and the CMCC-MED global model were used for this purpose. For GCMs, the A1B-driven simulations were considered in order to match the development trend assumed for RCM simulations at the middle between the extreme IPCC storylines. GCM simulations

Table 3.1 GCMs Used to Perturb Regional Model Outputs

Model	Res. (lat. × lon.)	Institution	Emission scenario	Acronym
HadCM3	2.5 × 3.75	UKMO (United Kingdom Meteorological Office)	A1B	UKMO
CGCM_2.3.2	2.8 × 2.8	MRI (Meteorological Research Institute)	A1B	MRI
CNRM_CM£	2.8 × 2.8	CNRM (Centre National de Recherches Météorologiques)	A1B	CNRM
CSIRO_Mk3.5	1.9 × 1.9	CSIRO (Commonwealth Scientific and Industrial Research Organization)	A1B	CSIRO
CCSM3	1.4 × 1.4	NCAR (National Center for Atmospheric Research)	A1B	NCAR
MIROC3.2	1.125 × 1.125	CCSR (Center for Climate System Research)	A1B	MIROC
GFDL_cm2.1	2.5 × 2	GFDL (Geophysical Fluid Dynamics Laboratory)	A1B	GFDL
ECHAM5	1.875 × 1.875	MPI (Max Planck Institute)	A1B	MPI
FGOALS	2.8125 × 2.8125	IAP (Institute of Atmospheric Physics)	A1B	IAP
CMCC-MED	0.75 × 0.75	CMCC (Euro-Mediterranean Center on Climate Change)	A1B	CMCC-MED

Note: GCM = General Circulation Model. Perturbations were carried out for 2006–65.

were selected based on skills at reproducing observed climate. Table 3.1 lists the GCMs chosen for the simulations.

To keep the computational load of the modeling within the study's time and resource constraints, different ranges of climate scenarios were used in the different components of the analysis (table 3.2). In some cases the full range of models (the RCM and the 10 perturbations) were used; in others, the two climate models representing the extremes of the spectrum were selected to represent the range of possible climate outcomes. Table 3.2 also illustrates the range of spatial scales used for the analysis: for example, AEZs for agriculture, sub-basins for water resources. Impact analysis was carried out by comparing values of the variables of interest expected to prevail, on average, in the medium and longer term (2006–35 and 2036–65), to a historical baseline (1976–2005).

Crop Modeling

This book analyzes the impacts of climate change on Nigeria's most important crops: sorghum, millet, maize, rice, cassava, and yams, which together account for about 80 percent of total value added in agriculture.

The DSSAT-CSM (Decision Support System for Agrotechnology Transfer— Cropping System Model), based on the modular approach described by Jones, Keating, and Porter (2001) and Porter, Jones, and Braga (2000), was applied (see appendix C for details) using precipitation, temperature, and solar radiation from climate projections, while keeping all the other input parameters fixed (e.g., crop management, soil, etc.). Although simulations were performed assuming both a fixed CO_2 concentration (380 ppm) and a higher concentration (582 ppm, consistent with the A1B emission scenario), in what follows only the results related to the fixed concentration are reported (appendix C provides the full set of results). As explained in box 3.3, there are still questions

Table 3.2 Treatment of Climate Model Uncertainty across the Areas of Analysis

Area of analysis	Variables modeled	Variation assessed	Climate scenarios used to define uncertainty band	Level of spatial disaggregation	Notes
Crop yield: impact analysis	Yield (t/ha)	% changes vs. 1976–2005 (rain-fed and irrigated)	Regional climate model (RCM) + 5 perturbations	Agro-ecological subzone (AESZ)	Selected perturbations on the basis of extreme climate conditions
Crop yield: adaptation analysis	Yield (t/ha)	% changes vs. 1976–2005 (rain-fed and irrigated)	RCM + 2 perturbations	AESZ	Selected perturbations on the basis of the two most extreme climate conditions
Livestock	Temperature-humidity index (THI) Gross primary productivity (GPP)	% changes vs. 1976–2005	RCM + 10 perturbations	From 8 km grid to AESZ	For GPP, just areas with grazing sustaining land covers (grassland, savanna)
Stream flow	Stream flow (mm)	% changes vs. 1976–2005	RCM + 10 perturbations	From 8 km grid to sub-basins	
Food security	Mean adequacy ratio (MAR) Nutrient adequacy ratio (NAR)	Changes in the absolute value	RCM + 2 perturbations	AESZ	MAR: Selected perturbations on the basis of the two most extreme climate conditions NAR: RCM, as one of the most pessimistic to assume precautionary conditions
Macroeconomic analysis	GDP, production, crop prices, imports, exports, net imports	% changes vs. baseline (i.e., future with no climate change)	RCM + 2 perturbations	Agro-ecological zone (AEZ)	Selected perturbations on the basis of the two most extreme climate conditions, matching analysis of impacts/adaptation on crops and food security
Irrigation	Reliability	Changes in the absolute value	RCM + 10 perturbations	One site	
Hydropower	Reliability of firm power supply; power production	Changes in the absolute value	RCM + 10 perturbations	Six sites	

about the ability of the CO_2 concentration effect to offset the negative impacts on yields of higher temperatures and more erratic precipitation. Model calibration and validation were undertaken in representative geographical locations to enable extrapolation of the results to the entire national territory.

The impact of climate on crop yields was analyzed using simulation results for a subensemble consisting of an RCM simulation and its five most extreme and significant GCM perturbations in terms of climate change projections. Yields obtained with weather data for the reference period 1976–2005 (baseline) were

Box 3.3 The CO_2 Fertilization Effect

The magnitude of the direct effect of increased CO_2 concentration on yield depends on the interaction of two mechanisms: the intensified photosynthesis activity and more efficient water use. In fact, a higher CO_2 concentration reduces stomatal aperture and stomatal density, which reduces stomatal conductance and thus transpiration (Olesen and Bindi 2002). An average reduction of 20 percent of stomatal conductance has been found when the CO_2 concentration is doubled (Drake, Gonzalez-Meler, and Long 1997), and CERES models included in DSSAT are able to consider this. It is also known that the CO_2 fertilization effect is usually stronger at higher temperatures (Goudriaan and Zadoks 1995); it seems that the highest effect should be found where growing season temperature increases are highest.

There are still uncertainties about the positive effects of higher atmospheric CO_2 concentration. Long *et al.* (2006) thought that crop models tend to overestimate the effect of CO_2 on plant growth and yield because CO_2-related model parameters are mainly derived from controlled and semi-controlled experiments, which typically show a higher CO_2 response than is seen under field conditions. Conversely, Tubiello *et al.* (2007) argued that, although as yet few experiments with free air carbon dioxide enrichment (FACE) have validated crop models under field conditions, there is growing evidence that crop models can reproduce the responses observed in the FACE experiments.

Independent of uncertainties linked to modeling processes at work under conditions of elevated CO_2, plant physiologists and modelers converge in affirming that the effects of elevated CO_2 may be overestimated by models because of such limiting factors as pests, weeds, nutrients, and other competition for resources that are neither well understood at large scale nor well represented in the models (Tubiello, Soussana, and Howden 2007).

Positive effects, reducing the negative impacts of a warmer climate and assuming increased atmospheric CO_2 concentration, were here recorded mainly for C3 species (see box 5.2): rice, cassava, and yams. Another important effect, for both C3 and C4 species, is the stomata closure, which improves water use efficiency, especially where precipitation is low, as in the northern AEZs. For the sake of clarity, C3 photosynthesis is the typical photosynthesis that most plants use, while C4 is an adaptation to arid conditions because plants need to use water more efficiently. However, the extent to which CO_2 regulates the effect of climate on crops is still debated, since this effect can be overestimated for many reasons and has not yet been fully proven experimentally.

compared with those obtainable under likely climate conditions over the medium and long term. The methodology is explained in detail in appendix C.

Food Security

Results for crop yields obtained with a reduced subensemble of climate projections consisting of RCM simulation and its two most extreme and significant GCM perturbations were integrated with average socioeconomic status (nutritional outcomes, demographic changes, and market access) to quantify future food security threats by AESZ.

Production of each of the major crops in 2000 stratified by AESZ was extracted from the Spatial Production Allocation Model (SPAM) dataset. The production values obtained were associated with yield and harvested areas for each AESZ. Variations of yield for each major crop type and demographic growth define both food crop availability and population requirements in 2000, 2020, and 2050.

Food security was analyzed by calculating the mean adequacy ratio (MAR) and nutrient adequacy ratio (NAR) for the baseline and future periods (see appendix D for details). MAR measures fulfillment of dietary requirements of energy for the population and nutrient intakes from the available food crop quantities. MAR is calculated by averaging the individual NARs (Hatløy, Torheim, and Oshaug 1998). NAR specific to calories or nutrients is defined as the ratio of energy or nutrients available per person from food crop quantities to the recommended nutrient intakes (RNI). NAR and MAR each equal 1 when average intake of energy and nutrients corresponds to the recommended intake; a lower average intake implies nutrient deficiency.

It was assumed, as a first approximation, that local demand (at the level of AEZs) can be met only by local production. Information on ease of access to markets (proxied by the travel time to the nearest market center) for each AESZ was used to assess the ability of the local population to reduce yield gaps (increase agricultural efficiency) and achieve food security, especially where NAR is below 1.

Livestock

For the livestock sector (see appendix E for details), two indicators of climate impacts were evaluated across the full range of climate models used:

- The temperature-humidity index (THI; Bohmanova, Misztal, and Cole 2007), selected as a proxy for the *suitability* to climate of livestock practices, was used to represent the climate (thermal) stress on livestock productivity, water requirements, and mortality.
- The natural gross primary productivity (GPP) of vegetation was used to represent livestock *sustainability* in a given territory, that is, the ability of the ecosystem under future climate conditions to generate sufficient animal feed to sustain livestock herds.

Details of the modeling approach applied to calculate both indicators are reported in appendix E. To summarize the total risk to livestock in terms of both thermal stress and reduced feed availability, findings from the two analyses were then consolidated into a single qualitative risk index. In each AESZ, percent changes in THI and GPP were averaged, classified into risk classes, and finally combined into a summary index. Results of both the individual consolidated analyses are reported in chapter 5.

Water Resources

The water resources analysis assessed the spatial and temporal availability of water resources for each HA in Nigeria so as to estimate the hydropower and irrigation potential at both current and planned small and large plants in selected watersheds.

Among numerous models used to simulate water balance at the watershed level, the Geographic Information System (GIS) version of the Soil and Water Assessment Tool (SWAT) model (ArcSWAT; http://swatmodel.tamu .edu/software/arcswat) was chosen to evaluate climate risk to water resources. ArcSWAT, which combines GIS and physical hydrological models, was found suitable (SWAT 2000, per Neitsch *et al.* 2002). After modeling the river network through a digital elevation model, 893 basins were defined for the physical hydrological analysis. Further, 234 soil types, 16 land covers, and 5 slope class layers were combined to extract hydrological response units (HRUs) that were assumed to have similar hydrological characteristics (for a detailed description of the model and how it was applied, see appendix F).

To analyze climate change impact on water resources, hydrological simulations of each of the 893 basins were made using the full ensemble of climate projection, consisting of RCM and the 10 GCM perturbations. Outputs were aggregated at 30-year intervals. The short- and medium-term results were compared with the baseline. The Niger and Benue main river stems were excluded from the analysis because their behavior depends on hydrological processes taking place outside Nigeria (and thus beyond the scope of the present study). In any case, water resource investments expected to take place along the main stems of the two rivers are marginal compared to developments planned in tributary basins.

In addition to the impacts at sub-basin scale, the book describes site-level case studies to assess climate impact on hydropower development and irrigation activities in both current and planned projects (see appendix H for a description). The assessment covers large, medium, and small dams; the single-purpose Shiroro, Zungeru, and Mambilla hydropower plants; and the multipurpose Gurara, Tiga, Dadinkowa, and Ikere Gorge schemes. The main purposes of the hydropower study were to assess the reliability of the power supply in future climate scenarios and to determine likely variations in total power and revenue generation. For irrigation schemes, the intent was to determine the optimal irrigation area in various climate scenarios.

For each site the RCM-simulated inflow for the baseline period, 1976–2005, was bias-corrected based on the historical record. The same coefficients were used to correct all the simulated inflows (RCM and its

GCM-based perturbations) for the future, 2006–65. The bias-corrected simulated inflows were used to run the energy generation model for the Shiroro, Zungeru, Gurara, Mambilla, Ikere, and Dadinkowa sites to calculate total power production and the reliability of industrial power during the historical period and for the future climate scenarios. The energy simulation was run monthly.

For the single-purpose plants, the objective criteria were to (a) maximize energy production at the commercial price of ₦6 per kilowatt hour (kWh), with a penalty of ₦6 per kWh for deficit power, and at a price of ₦2 per kWh for secondary power; and (b) minimize water spillage. For the multipurpose schemes at Gurara, Dadinkowa, and Ikere Gorge, the power model (see appendix H for details) was run to optimize energy production and minimize spill when water supply and irrigation needs were met. The irrigation assessment was conducted for the Tiga scheme to determine the optimal area for irrigation based on 80 percent assurance of the irrigation water requirement. The optimal areas for the historical period and future scenarios were compared.

Macroeconomic Analysis

The macroeconomic analysis evaluated the effects of climate-induced yield changes on macroeconomic outcomes (e.g., volume and composition of gross domestic product [GDP], imports/exports, etc.). The climate change impacts on agricultural production obtained from crop growth analysis were fed into a general equilibrium model, the Intertemporal Computable Equilibrium System (ICES). A preliminary step was to construct a future reference scenario that captured plausible economic development in Nigeria up to 2050 (table 3.3).

This reference scenario is the counterfactual "without climate change"; the impacts of climate change on crop productivity were imposed on it, and the consequent GDP and sectoral performance of the economic system were evaluated against it.

Assumptions for irrigation, consistent with the Master Plan for Irrigation and Dam Development but delayed by five years, are that in 2025 about 5 percent of Nigerian agriculture (2.1 million hectares [ha]) will be irrigated, to reach 25 percent of total agricultural land in 2050 (some 11 million ha). It is assumed that future yields will in relative terms be as vulnerable as current yields to climate shocks, so that deviations from current yields obtained from crop modeling can be applied to future yields. The rationale is that yield increase in the reference, no climate change, scenario will be achieved largely by expanding irrigation and through management practices suited to the current climate but not necessarily to a future warmer and more erratic climate. In particular, it is assumed that there will be minimal uptake of sustainable land management options like those discussed in chapter 6.

Table 3.3 Macroeconomic Assumptions, No Climate Change Reference Scenario

Period	Average GDP growth rate (%)
2010–20	9.0
2021–30	8.4
2031–40	6.0
2041–50	4.3
2010–25[a]	9.0
2025–50[a]	5.7
A. Sector shares in total value-added in 2025	*Vision 20: 2020* *Model simulation*
Agriculture	21% 23%
Manufacturing	18% 17%
Mining	15% 21%
Services	46% 39%
B. Agricultural productivity growth	
2010–18	3-fold 2.5-fold
2010–25	6-fold 5.3-fold
2010–50	— 19-fold

Note: — = not available.
a. These rates were calculated assuming that Vision 20: 2020 objectives are achieved with a five-year slippage.

Table 3.4 Agro-ecological Zoning Used in the Economic and Crop Models

Zones in the economic model	Zones in crop modeling
AEZ 1	AESZ 7
AEZ 2	AESZ 1 and AESZ 2
AEZ 3	AESZ 9 and AESZ 5
AEZ 4	AESZs 10, 8, 11, 4, 3
AEZ 5	AESZs 14, 12, 13, 6
AEZ 6	AESZ 15

Note: AEZ = agro-ecological zone; AESZ = agro-ecological subzone.

The model set up for this study contains several improvements compared to conventional computable general equilibrium (CGE) practice (see appendix I). The analysis was carried out for different climate models, representing the variability of yield changes—and correspondingly of macroeconomic impacts—across climate outcomes corresponding, on average,[1] to a less and a more pessimistic scenario of yield change.

Because of the structure of the social accounting matrix (SAM) used in the ICES model, crops and zones were disaggregated as follows: rice, cassava, and yams are modeled individually and millet, sorghum, and maize as a single aggregate, "other cereal crops." Six AEZs were analyzed—a slightly coarser disaggregation than the ones used for crop modeling. Table 3.4 shows the correspondence between the two levels of disaggregation.

Annex 3A Data Sources

Table 3A.1 Data Used as Inputs to the Analysis

Data	Units	Type	Source	Link	Where used in analysis	Notes
Data used for climate, hydrology, and agriculture analyses, including food security						
Precipitation	mm	Meteorological	Nigerian Meteorological Agency (NIMET) via World Bank		Climate, hydrology, and agriculture (crop and livestock) for calibration/ validation purposes	Diffuse spatiotemporal gaps; then station network to be integrated to better represent different landscape types and Nigerian subregions. Radiation daily data, used for crop model simulations, were calculated using CUP+ model, starting from CLIMWAT (FAO) monthly data.
Maximum temperature	°C	Meteorological	NIMET via WB			
Minimum temperature	°C	Meteorological	NIMET via WB			
Radiation	MJ/m²*day	Meteorological	National Climate Data Center (NCDC)	http://gis.ncdc.noaa.gov/map/cdo/?thm=themeDaily&layers=01	Hydrology and agriculture for calibration/ validation purposes	
Wind speed	m/s	Meteorological	NCDC			
Relative humidity	n.a.	Meteorological	NCDC		Hydrology and agriculture (crop and livestock) for calibration/ validation purposes	
Precipitation	mm	Meteorological	Climate Research Unit (CRU)	http://badc.nerc.ac.uk/view/ badc.nerc.ac.uk__ATOM__ dataent_1256223773328276	Climate for regional climate model (RCM) validation	Coarse resolution if compared to RCM; interpolation from stations often creates "smoothing"
Maximum temperature	°C	Meteorological	CRU			
Minimum temperature	°C	Meteorological	CRU			

table continues next page

Table 3A.1 Data Used as Inputs to the Analysis *(continued)*

Data	Units	Type	Source	Link	Where used in analysis	Notes
Discharge	m³/s	Hydrological	JICA MP (1995) via WB		Hydrology (site + country level)	
Evaporation	mm	Hydrological	JICA (1995) and dam feasibility studies		Hydrology (site level)	
Digital elevation model	masl	Topography	SRTM	http://www.cgiar-csi.org/data/srtm-90m-digital-elevation-database-v4-1	Hydrology and agriculture (crop)	Vertical accuracy ±15m
Slope	Degree	Topography	SRTM**			
Soil hydrologic group	n.a.	Soil	Harmonized World Soil Dataset (HWSD)	http://www.iiasa.ac.at/Research/LUC/External-World-soil-database/HTML/	Hydrology	Global dataset spatialized from soil profile data with different sources and dates; values derived from coarse scale interpolation procedures not fully suitable for soil data processing at national level
Maximum rooting depth of soil profile	mm	Soil	HWSD			
Soil layer depth	mm	Soil	HWSD		Hydrology and agriculture (crop)	
Bulk density	g/cm³	Soil	HWSD			
Available water content	mm H₂O/mm soil	Soil	HWSD		Hydrology	
Saturated hydraulic conductivity	mm/hr	Soil	HWSD			

table continues next page

Table 3A.1 Data Used as Inputs to the Analysis (continued)

Data	Units	Type	Source	Link	Where used in analysis	Notes
Organic carbon content	%	Soil	HWSD		Hydrology and agriculture (crop)	
Clay content	%	Soil	HWSD			
Silt content	%	Soil	HWSD			
Sand content	%	Soil	HWSD			
Rock fragment content	%	Soil	HWSD		Hydrology	
Soil albedo	n.a.	Soil	HWSD**	Ten Berge (1986)		
USLE (Universal Soil Loss Equation) soil erodibility	n.a.	Soil	HWSD**	RUSLE (revised USLE) science documentation; http://www .ars.usda.gov/Research/docs .htm?docid=6028		
pH in water	n.a.	Soil	HWSD	http://www.iiasa.acat/Research/ LUC/External-World-soil- database/HTML/	Agriculture (crop)	
Land cover 2006	Categories	Land cover	MOD12Q1	https://lpdaac.usgs.gov/products/ modis_products_table/mcd12q1	Hydrology and agriculture (livestock)	
Global land cover 2005–06	Categories	Land cover	GLC 2006; http:// ionia1.esrin.esa .int/		Hydrology	
Statistics on crop yields	t/ha	Crop	NPAFS-Report APS (2009)		Agriculture (crop)	

table continues next page

39

Table 3A.1 Data Used as Inputs to the Analysis (continued)

Data	Units	Type	Source	Link	Where used in analysis	Notes
Crop management	Planting and harvesting dates, fertilization, irrigation	Crop	USAID-MARKET; various bibliographic sources	http://www.nigeriamarkets.org/		
Cultivated areas (2000)	ha	Crop	Monfreda, Ramankutty, and Foley (2008), Ramankutty et al. (2008)	http://www.geog.mcgill.ca/~nramankutty/Datasets/Datasets.html	Agriculture (crop)	
Physical and harvested area, yields, and production in 2000	ha, t/ha, t	Food security	SPAM geodataset (You et al. 2010)	http://MapSPAM.info	Food security (crop)	
Calories and proteins	Calories (Kcals per 100 grams); protein (mg per gram)	Food security	FAOSTAT 2010	http://faostat.fao.org/	Food security (crop)	
Population density	Person/km^2	Food security	GRUMPv1 (CIESIN et al. 2011)		Food security (crop)	
Travel time to market centers	Hours	Food security	Travel time maps (Nelson 2008)		Food security (crop)	

table continues next page

Table 3A.1 Data Used as Inputs to the Analysis *(continued)*

Data	Units	Type	Source	Link	Where used in analysis	Notes
Data used in the economic analysis						
GDP base year (2004)	US$, millions	Economic	Nwafor, Diao, and Alpuerto (2010)		Used to build the economic baseline for Nigeria	For internal data consistency and availability the base year chosen is 2004, but corrected with International Food Policy Index (IFPRI) 2006 reporting quite different values for macrosectoral shares of value added.
GDP 2010–25	US$, millions	Economic	Nigeria Vision 20: 2020; the first national implementation plan	http://www.npc.gov.ng/home/doc.aspx?mCatID=68253	Used to build the economic baseline for Nigeria	
GDP 2026–50	US$, millions	Economic	Own assumption		Used to build the economic baseline for Nigeria	
National crop production 2004	US$, millions	Economic	Global Trade, Assistance, and Production database: Narayanan and Walmsley (2008)	http://www.gtap.agecon.purdue.edu/databases/v7/v7_doco.asp	Used to build the economic baseline for Nigeria	

table continues next page

41

Table 3A.1 Data Used as Inputs to the Analysis *(continued)*

Data	Units	Type	Source	Link	Where used in analysis	Notes
National crop production 2010–25	US$, millions	Economic	Endogenously produced by the model consistent with GDP and macro-sectoral shares target of the Nigeria Vision 20: 2020		Used to build the economic baseline for Nigeria	
National crop production 2026–50	US$, millions	Economic	Own assumption		Used to build the economic baseline for Nigeria	
Crop production per AESZ in the base year (2006)	tonnes and US$, millions	Economic	Extrapolated from Avetisyan, Baldos, and Hertel (2011)	http://www.gtap.agecon.purdue.edu/databases/v7/v7_doco.asp	Used to build the economic baseline for Nigeria	
Macrosectoral share of value added 2010–25	US$, millions	Policy, economic	Nigeria Vision 20: 2020; the first national implementation plan	http://www.npc.gov.ng/home/doc.aspx?mCatID=68253	Used to build the economic baseline for Nigeria	The economic analysis includes assumptions on the evolution of macro-sectoral shares as "scenario information" without including assumptions on the technologies, measures, and investment needed to meet that evolution. Therefore, no indication can be derived of the feasibility and costs of the different Nigeria Vision 20: 2020 targets.

table continues next page

Table 3A.1 Data Used as Inputs to the Analysis (*continued*)

Data	Units	Type	Source	Link	Where used in analysis	Notes
Increase in agricultural productivity 2010–25	% change compared to 2010	Policy, economic	Prudential interpretation of the Nigeria Vision 20: 2020: the first national implementation plan	http://www.npc.gov.ng/home/doc.aspx?mCatID=68253	Used to build the economic baseline for Nigeria	The economic analysis includes the assumptions on crop yield increase as "scenario information" without including assumptions on the technologies, measures, and investment needed to meet that increase. Therefore no indication can be derived of the feasibility and costs of the different Nigeria Vision 20: 2020 targets.
Increase in agricultural productivity 2026–50	% change compared to 2010	Policy, economic	Own assumption		Used to build the economic baseline for Nigeria	
Baseline expansion of irrigation 2010–50	Million hectares	Policy, economic	World Bank communication		Used to build the economic baseline for Nigeria and to compute the net impact of climate change on yields	

table continues next page

Table 3A.1 Data Used as Inputs to the Analysis *(continued)*

Data	Units	Type	Source	Link	Where used in analysis	Notes
Cost of irrigation per hectare	US$	Economic	You et al. (2009), integrated with World Bank direct communication		Used to compute the direct cost of additional irrigation needed to offset decline in yields in the climate change scenarios (with and without alternative adaptation measures) and its economic implications	
% change in yields	% change in tons produced per hectare	Crop	Mereu and Spano (2011)		Input to the economic assessment	Due to computational constraints, the change in yields considered for the economic analysis is associated with a subset of the full range of climate models. Therefore, the costs estimated have to be interpreted as a lower bound of the potential climate change impacts on agriculture.

Note: n.a. = not applicable.
Significance level: ** = Derived.

Note

1. The disclaimer "on average" is needed as there is no perturbed simulation providing the best or worst yield outcome in all AESZs and for all crops.

References

Araújo, M. B., and M. New. 2007. "Ensemble Forecasting of Species Distributions." *Trends in Ecology and Evolution* 22: 42–47.

Avetisyan, M., U. Baldos, and T. Hertel. 2011. "Development of the GTAP Version 7 Land Use Data Base." GTAP Research Memorandum No. 19, Center for Global Trade Analysis, Purdue University, West Lafayette, IN.

Bohmanova, J., I. Misztal, and J. B. Cole. 2007. "Temperature-Humidity Indices as Indicator of Milk Production Losses Due to Heat Stress." *Journal of Dairy Sciences* 90: 1947–56.

Buishand, T. A., and G. Lenderink. 2004. *Estimation of Future Discharges of the River Rhine in the SWURVE Project*. KNMI, Technical Report, Royal Netherlands Meteorological Institute, De Bilt, Netherlands.

CIESIN (Center for International Earth Science Information Network), IFPRI, the World Bank, CIAT. 2011. *Global Rural-Urban Mapping Project (GRUMPv1): Population Grids*. Palisades, NY: Socioeconomic Data and Applications Center (SEDAC), Columbia University.

Drake, B. G., M. A. Gonzalez-Meler, and S. P. Long. 1997. "More Efficient Plants: A Consequence of Rising Atmospheric CO_2?" *Annual Review of Plant Physiology and Plant Molecular Biology* 48: 609–39.

Goudriaan, J., and J. C. Zadoks. 1995. "Global Climate Change: Modeling the Potential Responses of Agro-Ecosystems with Special Reference to Crop Protection." *Environmental Pollution* 87: 215–24.

Haerter, J. O., S. Hagemann, C. Moseley, and C. Piani. 2011. "Climate Model Bias Correction and the Role of Timescales." *Hydrology and Earth System Sciences Discussion* 7: 7863–78. doi:10.5194/hessd-7-7863-2010.

Hatløy, A., L. E. Torheim, and A. Oshaug. 1998. "Food Variety—A Good Indicator of Nutritional Adequacy of the Diet? A Case Study from an Urban Area in Mali, West Africa." *European Journal of Clinical Nutrition* 52: 891–98.

JICA (Japan International Cooperation Agency). 1995. *The Study on the National Water Master Plan. Sector Report*. Vol 2. Report prepared for the Federal Ministry of Water Resources and Rural Development, Abuja.

Jones, J. W., B. A. Keating, and C. H. Porter. 2001. "Approaches to Modular Model Development." *Agricultural Systems* 70: 421–43.

Lionello, P., M. Gacic, D. Gomis, R. Garcia-Herrera, F. Giorgi, F. Planton, R. Trigo, A. Theocharis, M. N. Tsimplis, U. Ulbrich, and E. Xoplaki. 2012. "Program Focuses on Climate of the Mediterranean Region." *Eos, Transactions of the American Geophysical Union* 93 (10): 105. doi:10.1029/2012EO100001.

Long, S. P., E. A. Ainsworth, A. D. B. Leakey, J. Nosberger, and D. R. Ort. 2006. "Food for Thought: Lower-than-Expected Crop Yield Stimulation with Rising CO_2 Concentrations." *Science* 312: 1918–21.

Mereu, V., and D. Spano. 2011. *Climate Change Impacts on Crop Production*. Report for the World Bank tender, Climate Risk Analysis over Nigeria, Euro-Mediterranean Center on Climate Change, Lecce, Italy.

Monfreda, C., N. Ramankutty, and J. A. Foley. 2008. "Farming the Planet. Part 2: The Geographic Distribution of Crop Areas and Yields in the Year 2000." *Global Biogeochemical Cycles* 22: GB1022. doi:10.1029/2007GB002947.

Narayanan, G. B., and T. L. Walmsley, eds. 2008. "Global Trade, Assistance, and Production: The GTAP 7 Data Base." Center for Global Trade Analysis, Purdue University, Lafayette, IN.

Neitsch, S. L., J. G. Arnold, J. P. Kiniry, J. R. Williams, and K. W. King. 2002. *Soil and Water Assessment Tool (SWAT) Theoretical Documentation, Version 2000*. Temple, TX: USDA Agricultural Research Service.

Nelson, A. 2008. "Travel Time to Major Cities: A Global Map of Accessibility." Global Environment Monitoring Unit–Joint Research Centre of the European Commission, Ispra, Italy.

Nwafor, M., X. Diao, and V. Alpuerto. 2010. *A 2006 Social Accounting Matrix for Nigeria: Methodology and Results*. Report NSSP007, International Food Policy Research Institute, Washington, DC.

Olesen, J. E., and M. Bindi. 2002. "Consequences of Climate Change for European Agricultural Productivity, Land Use and Policy." *European Journal of Agronomy* 16: 239–62.

Olesen, J. E., T. R. Carter, C. H. Diaz-Ambrona, S. Fronzek, T. Heidmann, T. Hickler, T. Holt, M. I. Minguez, F. Morales, J. Palutikof, M. Quemada, M. Ruiz-Ramos, G. Rubik, F. Sau, B. Smith, and M. Sykes. 2007. "Uncertainties in Projected Impacts of Climate Change on European Agriculture and Terrestrial Ecosystems Based on Scenarios from Regional Climate Models." *Climatic Change* 81: 123–43.

Porter, C., J. W. Jones, and R. Braga. 2000. "An Approach for Modular Crop Model Development." International Consortium for Agricultural Systems Applications, Honolulu, HI, p. 13. Available from http://icasa.net/modular/index.html.

Ramankutty, N., A. Evan, C. Monfreda, and J. A. Foley. 2008. "Farming the Planet. Part 1: The Geographic Distribution of Global Agricultural Lands in the Year 2000." *Global Biogeochemical Cycles* 22: GB1022. doi:10.1029/2007GB002952.

Rockel, B., A. Will, and A. Hense. 2008. "The Regional Climate Model COSMO-CLM (CCLM)." *Meteorologische Zeitschrift* 17 (4): 347–48.

Scoccimarro, E., S. Gualdi, A. Bellucci, A. Sanna, P. G. Fogli, E. Manzini, M. Vichi, P. Oddo, and A. Navarra. 2011. "Effects of Tropical Cyclones on Ocean Heat Transport in a High Resolution Coupled General Circulation Model." *Journal of Climate* 24: 4368–84.

Smith, J. B., H.-J. Schellnhuber, M. Qader Mirza, S. Fankhauser, R. Leemans, L. Erda, L. A. Ogallo, B. A. Pittock, R. Richels, C. Rosenzweig, U. Safriel, R. S. J. Tol, J. Weyant, and G. Yohe. 2001. "Vulnerability to Climate Change and Reasons for Concern: A Synthesis." In *Climate Change 2001: Impacts, Adaptation, and Vulnerability*, edited by J. J. McCarthy, O. F. Canziani, N. A. Leary, D. J. Dokken, and K. S. White, 913–67. Cambridge, U.K.: Cambridge University Press.

Ten Berge, H. F. M. 1986. "Heat and Water Transfer at the Bare Soil Surface: Aspects Affecting Thermal Imagery." PhD thesis, Agricultural University Wageningen, The Netherlands.

Tubiello, F. N., J. S. Amthor, K. J. Boot, M. Donatelli, W. Easterling, G. Fischer, R. M. Giord, M. Howden, J. Reilly, and C. Rosenzweig. 2007. "Crop Response to Elevated CO_2 and World Food Supply: A Comment on 'Food for Thought,'" edited by Long *et al. Science* 312: 1918–1921, 2006. *European Journal of Agronomy* 26: 215–23.

Tubiello, F. N., J. F. Soussana, and S. M. Howden. 2007. "Crop and Pasture Response to Climate Change." *Proceedings of the National Academy of Sciences of the United States of America* 104 (50): 19686–90.

You, L., C. Ringler, G. Nelson, U. Wood-Sichra, R. Robertson, S. Wood, G. Zhe, T. Zhu, and Y. Sun. 2009. "Torrents and Trickles: Irrigation Spending Needs in Africa." Background Paper 9, African Infrastructure Country Diagnostic, World Bank Group, Washington, DC.

You, L., S. Crespo, Z. Guo, J. Koo, W. Ojo, K. Sebastian, M. T. Tenorio, S. Wood, and U. Wood-Sichra. Spatial Production Allocation Model (SPAM) 2000 Version 3 Release 2. http://MapSPAM.info (accessed August 31, 2011).

CHAPTER 4

Climate Projections

Nigeria's Climate: Features and Trends

Rainfall in Nigeria is driven by the seasonal migration of the intertropical convergence zone, where hot and dry easterly winds from the Sahara meet humid air from the Atlantic. The climate is semi-arid in the north and humid in the south; wet and dry seasons are distinct. The rainy season varies between three and seven months from the northeast of Nigeria to the south. Mean annual rainfall nationwide is estimated at 1,150 mm—about 1,000 mm in the center of the country, 500 mm in the northeast, and up to 3,500 mm along the coasts.

Ayoade (1970, 1973) reported that the southern zone showed no trend in precipitation in the middle of the 20th century, but a decade later a nation-wide study highlighted a reduction in precipitation during the second half of the century. Aina and Adejuwon (1995) found that rainfall was decreasing in the tropical area; they considered the main cause to be the removal of forests, which recycle water through evapotranspiration to the atmosphere. NIMET (Nigerian Meteorological Agency) data from 1941 to 2000 suggest that rainfall trends are more spatially heterogeneous than temperature trends (BNRCC 2011).

The overall features of the future global climate are becoming clearer, but regional patterns are still uncertain. More specifically, while the warming dynamics are relatively uncontroversial both globally and regionally, precipitation patterns, evapotranspiration, soil moisture, and runoff are not.

The recent National Adaptation Strategy and Plan of Action on Climate Change for Nigeria (NASPA-CCN) (BNRCC 2011) confirmed that although annual rainfall increases were projected in some parts of the country and decreases in others, all areas show rainfall increasing during at least part of the year. The same study also predicted a general increase in both rainfall and temperature extremes but with more uncertainty where rainfall was concerned.

Projecting Change

Regional climate model (RCM) characteristics, including how well they represent the current climate, are extensively discussed in appendix B. The results obtained through analyses of perturbed temperature and precipitation data are presented in the next sections.

Air Surface Temperature Projections

The simulated air surface temperature averaged over Nigeria (figure 4.1) shows a definite increasing trend. At the beginning of the 21st century climate projections ranged from 27.2°C to 27.6°C; the 0.4°C spread by the end of the simulation period may become 1.2°C (28.8°C–30.0°C). Taking into account that climate change models are uncertain, as illustrated by the perturbed results,

Figure 4.1 Air Surface Temperatures Averaged, 1976–2065

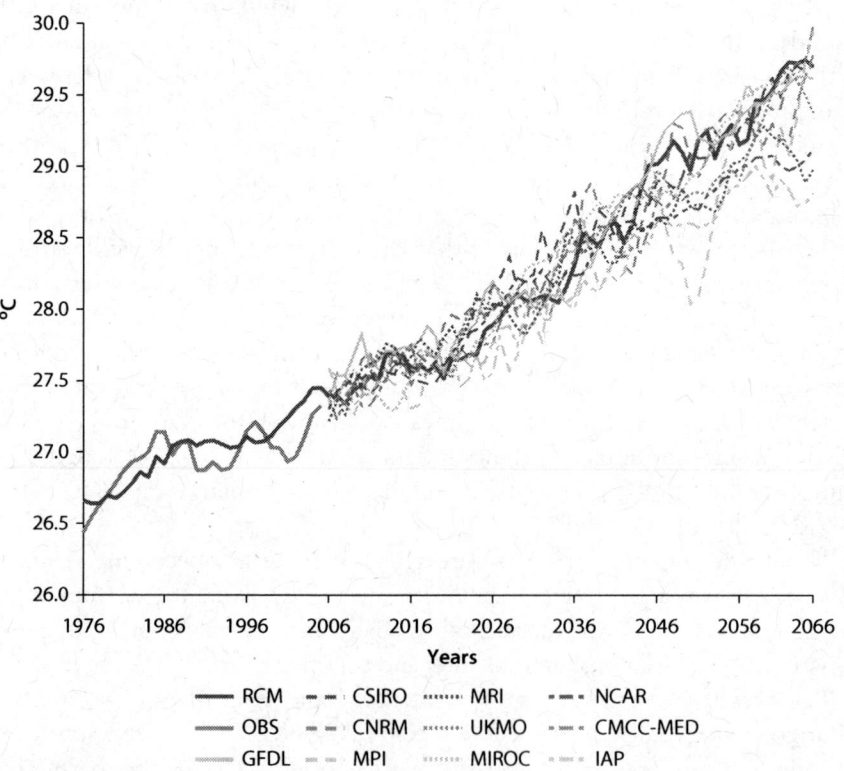

Source: Authors' calculations based on data sources listed in table 3A.1.
Note: Trends represent observations (solid orange line) and simulations using the RCM (solid blue line) and its GCM-based perturbations from 2006. OBS = observations; RCM = Regional Climate Model; CMCC-MED = Euro-Mediterranean Center on Climate Change; CNRM = Centre National de Recherches Météorologiques; CSIRO = Commonwealth Scientific and Industrial Research Organization; GFDL = Geophysical Fluid Dynamics Laboratory; IAP = Institute of Atmospheric Physics; MIROC = Center for Climate System Research; MPI = Max Planck Institute; MRI = Meteorological Research Institute; NCAR = National Center for Atmospheric Research; UKMO = United Kingdom Meteorological Office.

figure 4.1 indicates that average temperatures in Nigeria will be 1–2 degrees higher in 2050 than they are at present.

Using +1°C and +2°C (compared to historical patterns) as limits to classify range boundaries,[1] in the short-term future, the entire country seems to be seeing a moderate rise in surface air temperature. The map of classes (map 4.1) in terms of temperature variation for the medium term shows that the increase in temperature is higher toward the North. In the map the spatial disaggregation is by agro-ecological subzone (AESZ), which is the unit used for the crop modeling (appendix C).

Both the high-resolution model (RCM) and the perturbed Global Circulation Model (GCM) projections indicate similar spatial and temporal variations in temperature changes. The results of the high-resolution model (map 4.2) illustrate the difference in change between seasons and regions of the country. The warming projected for 2056–65 compared to 2001–10 is more evident in December to February (map 4.2, panel a), when the central part (from 7 to 12°N) of Nigeria is affected by warming of up to 3.5°C. From June through August of 2056–65 (map 4.2, panel b) warming is less pronounced; it reaches 2.8°C in the northern part.

Analysis of extreme events, performed exclusively at regional high resolution via RCM, suggests tendencies to increase for both extremely low and extremely high temperature values. The southern part of Nigeria (south of 7°N) is likely to be less affected by there being more extremely high temperature events.

Map 4.1 Distribution of Temperature Increase in 2050 Compared to 1990

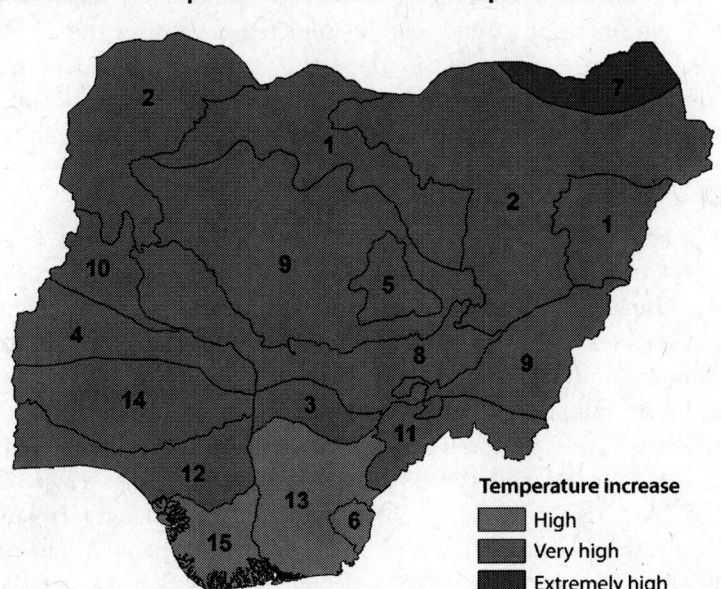

Temperature increase
- High
- Very high
- Extremely high

Source: Authors' calculations based on data sources listed in table 3A.1.
Note: Numbers refer to agro-ecological subzones (AESZs).

Map 4.2 Air Surface Temperature Increases in 2056–65 Compared to 2001–10

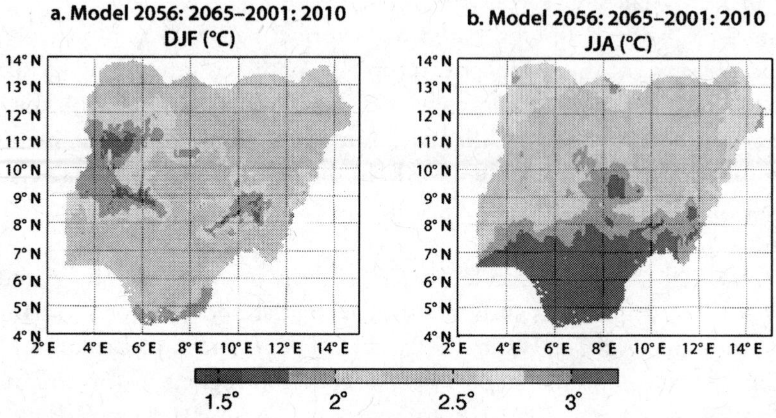

a. Model 2056: 2065–2001: 2010
DJF (°C)

b. Model 2056: 2065–2001: 2010
JJA (°C)

1.5°   2°   2.5°   3°

Source: Authors' calculations based on data sources listed in table 3A.1.
Note: Panel a: differences during the winter season (December, January, February); panel b: differences during the summer season (June, July, August).

In winter, the increasing trend in absolute values of extreme low temperature events is more pronounced than the changes in extreme high temperature events. The opposite is found in summer. All trends are more pronounced toward the end of the simulation period.

Surface Precipitation Projections

Figure 4.2 shows the precipitation time series averaged for all of Nigeria for 1976–2065. No significant trends can be detected in most of the 2001–65 perturbed precipitation time series obtained through GCM-based perturbations; only the data perturbed through the GFDL model shows a significant negative trend. The perturbed precipitation averaged over Nigeria at the beginning of the 21st century ranges from 3 to 3.5 mm/d with a spread of 0.5 mm/d, which becomes 1.4 mm/d at the end of the century.

Since climate models tend to disagree about how much change in precipitation to expect, the model results were summarized by defining four classes of risk (table 4.1). Using ± 15 percent as the stability band of percent changes from historical averages, a given sub-basin is considered stable if most climate models (those falling within the range of the 1st to the 99th percentiles of the ensemble) agree that future rainfall will not be more or less than 15 percent of historical values. Sub-basins are considered exposed to "dry risk" if the 1st percentile is less than minus 15 percent and the 99th percentile is less than 15 percent, and exposed to "wet risk" when the 99th percentile of changes is larger than 15 percent but the first percentile is more than minus 15 percent. The projection is considered uncertain when both a decline and an increase of more than 15 percent are considered possible.

It was found (see map 4.3) that around 2020 conditions in 53 percent of Nigeria's area are expected to be wetter, 10 percent will have less rain, 35 percent

Figure 4.2 Annual Precipitation over Nigeria, Averaged, 1976–2065

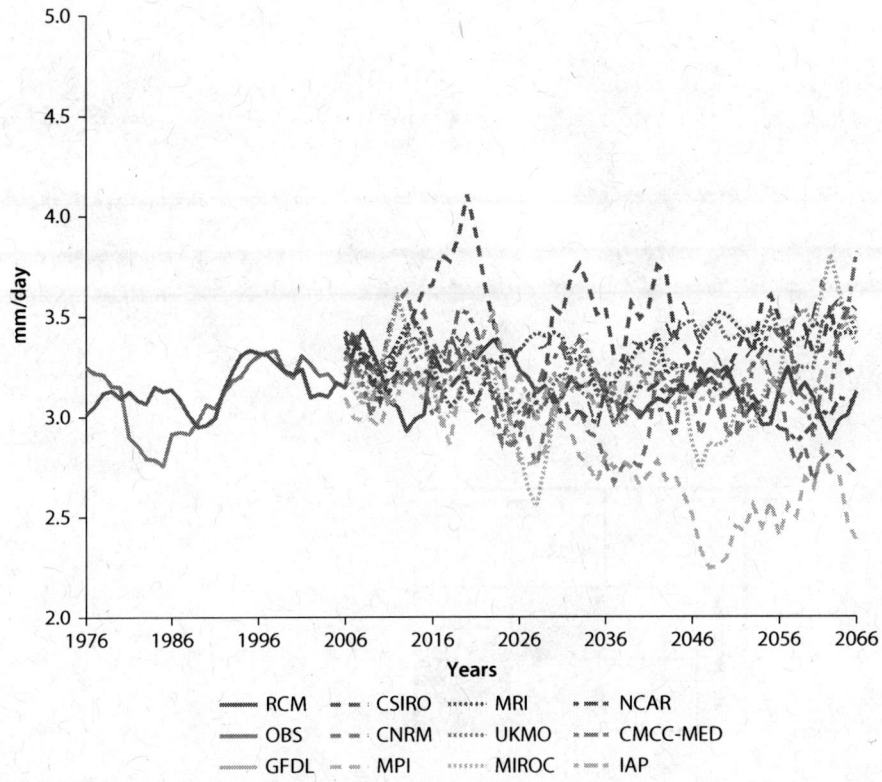

Source: Authors' calculations based on data sources listed in table 3A.1.
Note: Trends represent observations (solid orange line) and simulations using the RCM (solid blue line) and its GCM-based perturbations from 2006. OBS = observations; RCM = Regional Climate Model; CMCC-MED = Euro-Mediterranean Center on Climate Change; CNRM = Centre National de Recherches Météorologiques; CSIRO = Commonwealth Scientific and Industrial Research Organization; GFDL = Geophysical Fluid Dynamics Laboratory; IAP = Institute of Atmospheric Physics; MIROC = Center for Climate System Research; MPI = Max Planck Institute; MRI = Meteorological Research Institute; NCAR = National Center for Atmospheric Research; UKMO = United Kingdom Meteorological Office.

Table 4.1 Risk Classes of Changes in Water Flows from Historical Averages
percent

Risk class	1st percentile	99th percentile
Stable	>−15	<15
Dry risk	<−15	<15
Wet risk	>−15	>15
Uncertain	<−15	>15

will be stable, and for the remaining 2 percent precipitation projections are highly uncertain. In 2050, 41 percent of the country is expected to be wetter, 14 percent drier, and 20 percent stable, but the area subject to uncertainty increases from 2 to 25 percent. Evident clusters of drying areas in the short and medium term are concentrated in the southeast plateau and along the southwest

Map 4.3 Distribution of Classes of Precipitation Changes

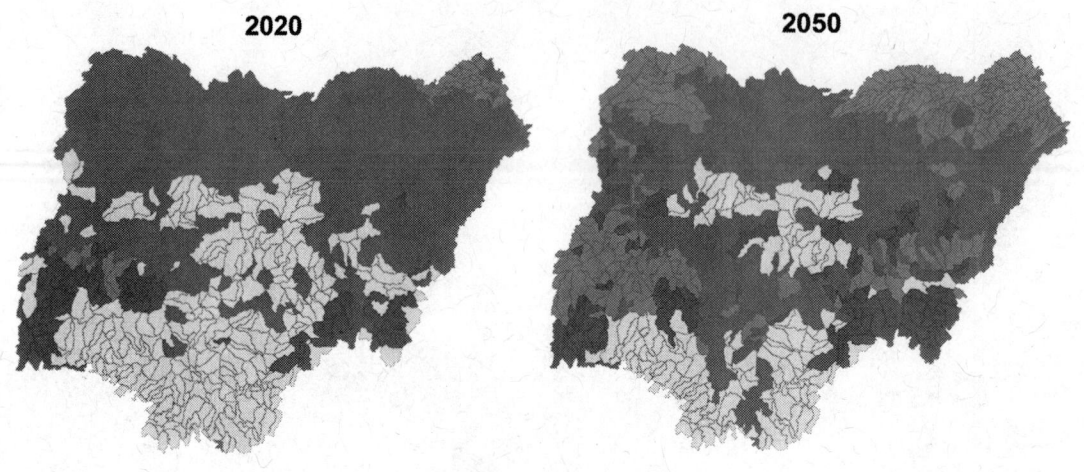

		1st percentile	
		< -15%	**> -15%**
99th percentile	**< 15%**	Dry risk	Stable
	> 15%	Uncertain	Wet risk

Source: Authors' calculations based on data sources listed in table 3A.1.
Note: Distribution of classes of precipitation changes in 2020 (left) and 2050 (right) compared to 1990. Spatial representation is in sub-basins, the units on which the hydrological (strictly rain-dependent) analysis was based.

littoral, with stable areas in the center and along the central and eastern coastal zones. Wetting areas are in the north, and uncertainty is evident mainly in the arid/semi-arid regions in the medium term.

Note

1. Temperature increase classification: low: 10th percentile < 1°C and 90th percentile < 1°C; moderate: 10th percentile < 1°C and 1°C < 90th percentile < 2°C; high: 1°C < 10th percentile < 2°C and 1°C < 90th percentile < 2°C; very high: 1°C < 10th percentile < 2°C and 90th percentile > 2°C; extremely high: 10th percentile > 2°C and 90th percentile > 2°C.

References

Aina, E. O., and S. A. Adejuwon. 1995. "Regional Climate Change: Implication on Energy Production in the Tropical Environment." In *Proceedings of the International Workshop on Global Change Impact on Energy Development,* edited by J. C. Umolu. Lagos, Nigeria: Damtech Nigeria Limited.

Ayoade, J. O. 1970. "The Seasonal Incidence of Rainfall in Nigeria." *Weather* 25: 414–18.

———. 1973. "Trends and Periodicities in Annual Rainfall in Nigeria." *Nigeria Geography Journal* 16 (2): 167–76.

BNRCC (Building Nigeria's Response to Climate Change). 2011. *National Adaptation Strategy and Plan of Action on Climate Change for Nigeria* (NASPA-CCN). http://nigeriaclimatechange.org/docs/naspaAug2012.

CHAPTER 5

Climate Change Impact Analysis

Crop Yields

To estimate changes in yield for the main Nigerian crops, climate projections from the regional climate model (RCM) and its perturbations were used to run crop models. Climate impacts proved to vary considerably by agro-ecological subzone (AESZ) and crop type. The differences in impacts on yield relate to the sensitivity of the specific crop to changed climate conditions and to crop distribution and the crop calendar. The impacts tend to increase between the short and the medium term. For reasons already explained (see in particular box 3.3), only the results based on a fixed CO_2 concentration are reported here. The full set of results, including increases in CO_2 atmosphere concentration, is provided in appendix C.

In this section, results are aggregated across agro-ecological zones (AEZs) to identify impacts at the level of individual crops, and across crops to produce results at the AEZ level, using base-year information on production shares and value added to define weights used for aggregating. Clearly, these weights will vary over time, since farms will respond to evolving biophysical, climatic, and economic drivers by modifying cropping patterns (subject to constraints in terms of technology, resource endowments, and local suitability of crops). Thus, the comparative static analysis presented here will be complemented by the dynamic analysis discussed later in the chapter, where the evaluation of impacts will take into account changes in cropping patterns and value added that are likely to take place in response to both climatic and economic drivers.

In terms of impacts on crops, results for the longer term (2050) show lower yields, with negative median values for all crops in 2050 (figure 5.1). However, the outlook for yams, millet, and cassava is uncertain, particularly in 2020 (figure 5.2), where the median for the climate models indicates the possibility of mild yield increases (3–6 percent or less). In 2050 the consensus of models is clearer, with 70 percent pointing to lower yields. Rice appears to be the most vulnerable crop throughout, with yields falling as much as 7 percent in the short term and 25 percent in the longer term.

Figure 5.1 Aggregate Percent Change in Crop Yields, 2050

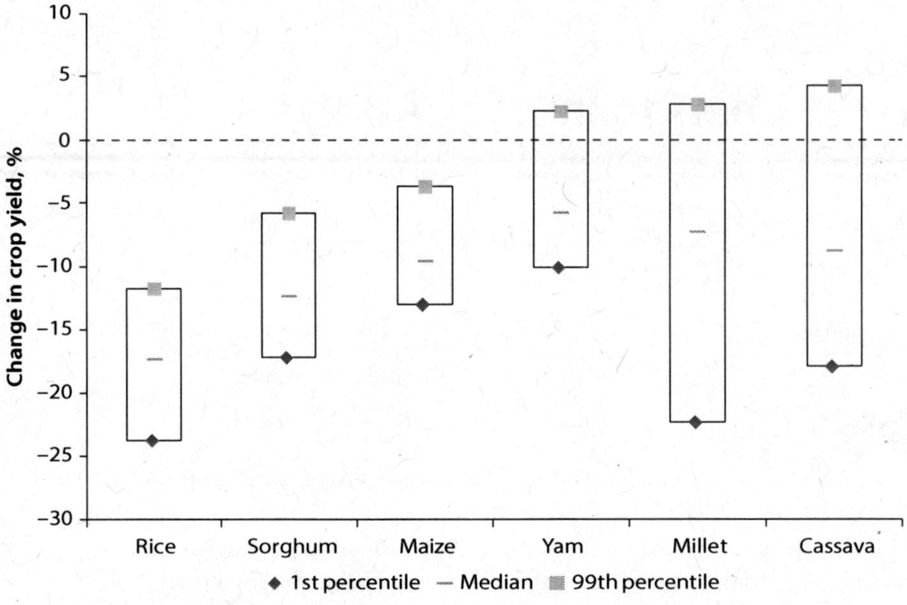

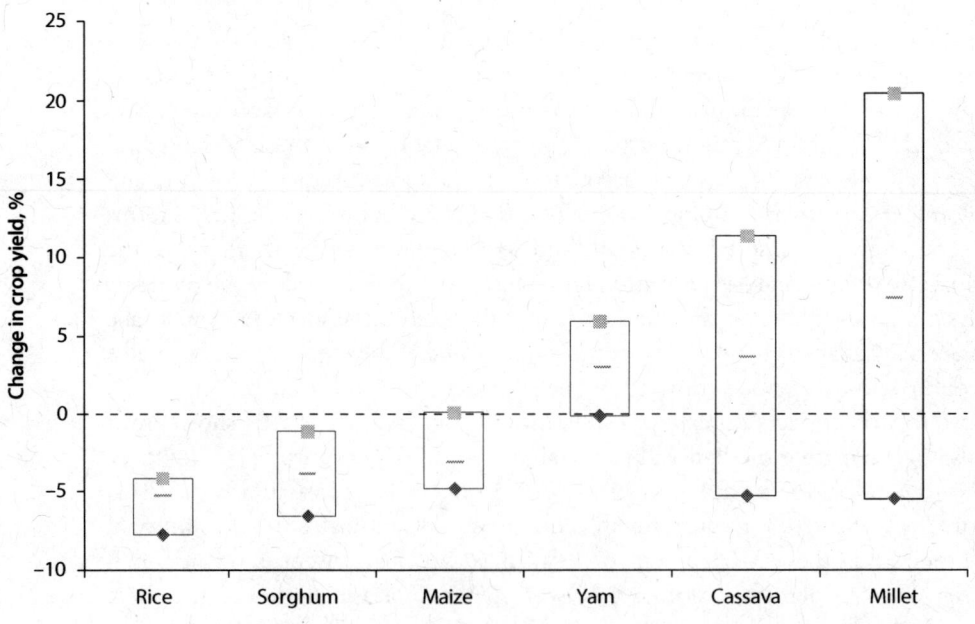

Source: Authors' calculations based on data sources listed in table 3A.1.
Note: For each crop, yield changes are aggregated across agro-ecological subzones (AESZs) and weighted by using base year share in crop production.

Figure 5.2 Aggregate Percent Change in Crop Yields, 2020

Source: Authors' calculations based on data sources listed in table 3A.1.
Note: For each crop, yield changes are aggregated across agro-ecological subzones (AESZs) and weighted by base year share in crop production.

Temperature change is likely to be the major driver of yield shocks, rather than precipitation (this is consistent with several recent studies, such as Lobell *et al.* 2008 and Lobell and Burke 2010). While rainfall variability is unquestionably important in driving year-to-year yield, there is widespread appreciation of the fact that a strong signal of temperature increase above the historical range (consistent across climate models) is likely to be the main driver of yield change, particularly when signals related to precipitation change are less clear. Temperature increase shortens the crop-growing period and reduces the amount of biomass that accumulates. This suppresses crop yields even if crops are not stressed by water conditions. It has been shown that there is no linear correlation between yield and water (Jones and Thornton 2003; Thornton *et al.* 2009) or with other nonclimatic factors affecting crop production (e.g., soil characteristics, management options).

In terms of impacts at the AESZ level, obtained by aggregating impacts on individual crops using nationwide crop shares in value added as weights, AESZs in the North (figure 5.3) appear more subject to risks of large declines (close to 20 percent in 2020 and 40 percent in 2050), but there is even more uncertainty, with yield increases in the more optimistic model up to 20 percent in 2020 and almost 10 percent in 2050. Despite the significant amount of variability across space, by 2050 aggregate yield decline seems more likely in all zones, as indicated by the negative median values observed in the lower part of figure 5.3.

In terms of results at the level of both crops and AESZ, full results are reported in appendix C; key findings are that

- In the short term (2006–35) in some AESZs, cereals show yield reductions, which accelerate in the medium term (2036–65).

Figure 5.3 Aggregate Percent Change in Crop Yields by AESZ, 2020 and 2050

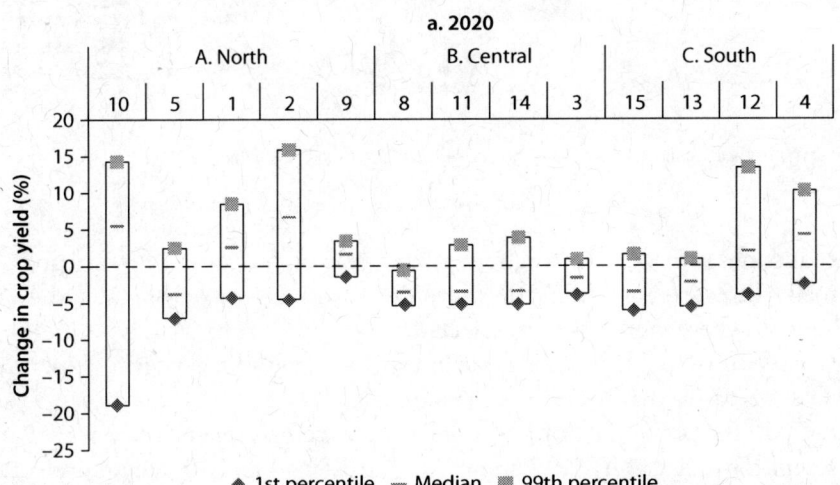

figure continues next page

Figure 5.3 Aggregate Percent Change in Crop Yields by AESZ, 2020 and 2050 *(continued)*

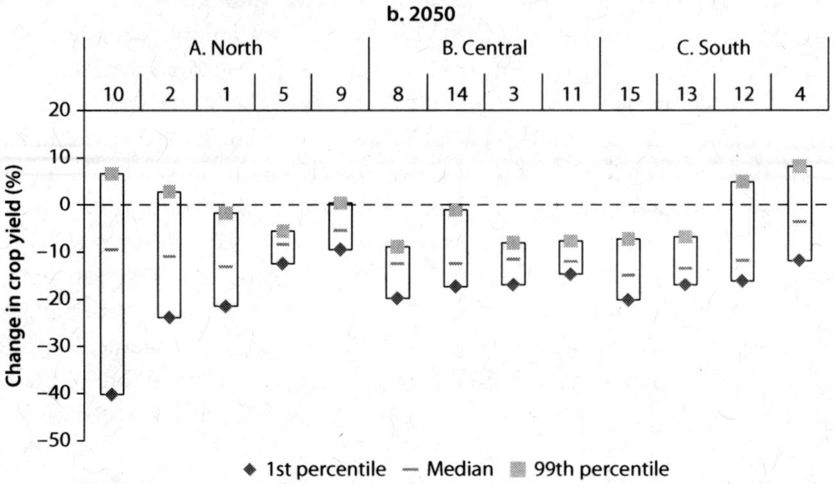

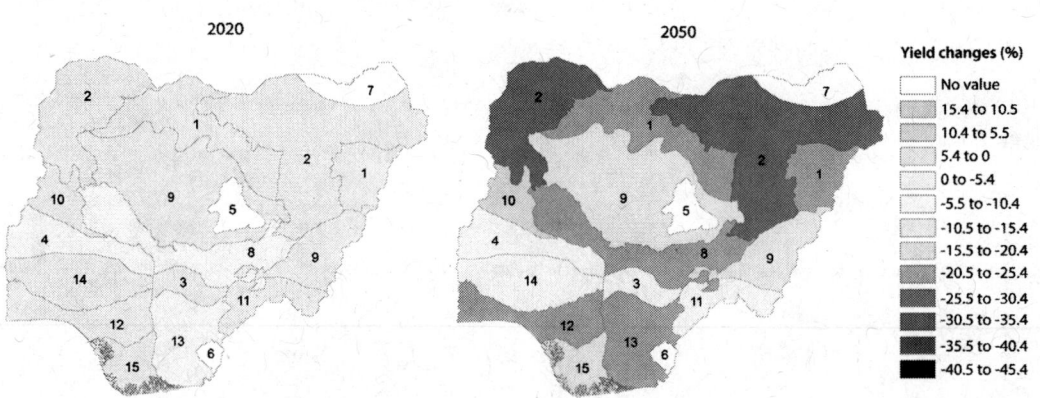

Source: Authors' calculations based on data sources listed in table 3A.1.
Note: AESZ = agro-ecological subzone. Yield changes are aggregated across crops using base year shares of value-added as weights.

Map 5.1 Changes in Rice Yields (Intermediate Model)

Source: Authors' calculations based on data sources listed in table 3A.1.
Note: The figure reports results for a climate model roughly representing the median of all the models. Numbered units refer to agro-ecological subzones.

- Reductions projected for sorghum, millet, maize, and rice in 2020 are probable in all AESZs except AESZ 10, where the uncertainty is very high, and AESZs 1 and 2, where increases are projected.
- Rice seems particularly vulnerable in the north, with longer-term reduction in yields of 20–30 percent or more (map 5.1).
- By 2050 the probability of lower yields in all cereals in all AESZs is very high except for millet in AESZ 2 and maize in AESZs 4 and 10, where projections are clouded by uncertainty.

- Results for the root crops, cassava and yams, show high variability in yield impacts: the more pessimistic models suggest yield decline in all AESZs in both 2020 and 2050, while more optimistic models in some cases show significant increases in cassava yield (e.g., in AESZs 4, 9, and 12) for both.
- Impacts on the yam crop suggested lower yield increases in some AESZs in 2020, with high concordance between models. Decreases are projected generally for 2050, even though increases are projected for some AESZs.

Food Security

Food security outcomes depend not only on the net balance of food available nationally (including domestic production and imports, minus exports and stock accumulation), but also on how speedily and cost-effectively the food available can be moved around the country or stored for Nigerians in need. This in turn depends on, among other things, the quality of the transport infrastructure and the existence and reliability of food storage facilities.

While a full-blown analysis of how these factors interact with climate drivers to determine the degree of food security is beyond the scope of this book, it is of interest to examine how changes in local production affect food security, assuming, as a first approximation, that local demand in AEZs can be met only by local production. Worsening food security on this assumption can be interpreted as an indication of the heightened need for increased imports, or for improved functioning of national markets (including better transport and storage infrastructure), or for both.

An evaluation of mean adequacy ratios[1] (MAR) for the base year (2000 in this case) suggests that while current food security conditions are adequate (the national index exceeds the safety threshold of 1), six AESZs, accounting for about 50 percent of the population, have MAR values less than 1. With climate change (table 5.1), the situation deteriorates: the number of areas with inadequate MAR increase to 7 in 2020 and 11 in 2050, and at the national level the MAR worsens some 30 percent in 2020 and 70 percent in 2050. The spatial distribution of changes in the MAR is illustrated in map 5.2.

Livestock

Risks Related to Thermal Stress

As explained in chapter 3 a temperature-humidity index (THI) was calculated to evaluate changes in climatic conditions leading to possible heat stress for livestock. Using 75 as a value of THI as a safety threshold (no discomfort), it was found that in general the THI is likely to increase across Nigeria. The south is already on alert ($75 \leq THI < 79$) and will soon be moving into the danger zone ($79 \leq THI < 84$), while the north will shift from discomfort ($72 \leq THI < 75$) toward danger.

Table 5.1 Mean Adequacy Ratio (MAR) by AESZ, Year, and Climate Model

Agro-ecological subzone	Baseline (2000)	Low climate impact (2020)	High climate impact (2020)	Low climate impact (2050)	High climate impact (2050)
Dry sub-humid high plain	1.22	0.86	0.80	0.36	0.34
Dry sub-humid plain	1.94	1.50	1.38	0.70	0.66
Humid lowland and scarpland	2.08	1.55	1.50	0.86	0.80
Humid plain	1.90	1.18	1.13	0.46	0.42
Humid plateau	0.91	0.55	0.56	0.24	0.24
Perhumid high plain	0.65	0.53	0.50	0.33	0.31
Semi-arid plain	4.85	3.99	3.61	2.26	2.16
Sub-humid central Niger-Benue trough	4.07	2.93	2.83	1.52	1.42
Sub-humid high plain	2.51	1.66	1.65	0.77	0.76
Sub-humid plain	5.14	3.88	3.90	2.44	2.45
Very humid high plain	4.81	3.91	3.59	2.46	2.24
Very humid lowland	0.40	0.22	0.21	0.07	0.07
Very humid lowland and scarpland	0.98	0.68	0.64	0.31	0.28
Very humid plain	0.46	0.26	0.26	0.09	0.09
Very humid/perhumid Niger delta	0.58	0.37	0.38	0.17	0.17
Total Nigeria					
MAR	1.41	0.93	0.89	0.38	0.36
Change from 2000	0%	−34%	−37%	−73%	−74%

Source: Authors' calculations based on data sources listed in table 3A.1.
Note: The MAR is calculated for current conditions in 2000 and projected for 2020 and 2050 according to a low- and high-impact climate change model.

To capture the variability of THI across climate models, classes of risk were defined (table 5.2) as a function of agreement among climate models on the magnitude of change over time. The consensus band was defined by the 10th and 90th percentile in the climate model distribution.

The results are presented in map 5.3, which shows significant consensus on very high increases in much of the south, high increases in the center and the north, and moderate increases in the center-north.

Changes in THI indicate shifts in relative risks faced by different zones in terms of thermal stress; since there are no Nigeria-specific studies on linkages between thermal stress and livestock health, it is difficult to draw conclusions about the implications for mortality or morbidity rates; however, studies in Europe (box 5.1) suggest there is a significant correlation, which means these findings have policy significance and highlight the need to better quantify the impacts.

The second pathway of climate impacts on livestock evaluated in this book is their effect on the availability of feed, proxied by gross primary productivity (GPP). As described in appendix E, regression analysis was used to establish a relationship between climate variables and seasonal GPP, averaged over historical (1976–2005) and future (2006–35 and 2036–65) time periods.

Map 5.2 Percentage MAR Reduction for Each AESZ between 2000 and 2050

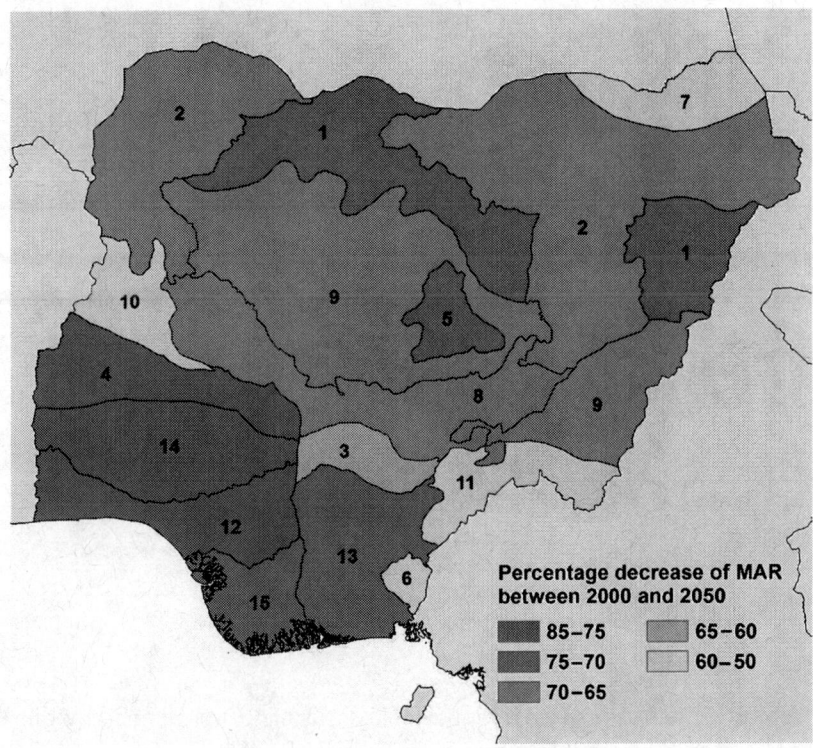

Source: Authors' calculations based on data sources listed in table 3A.1.
Note: AESZ = agro-ecological subzone; MAR = mean adequacy ratio. The percentage MAR reduction for each AESZ between 2000 and 2050 (MAR in 2000 minus MAR in 2050 over MAR in 2000) is calculated using the higher-impact climate model.

Table 5.2 Risk Classes for Higher Thermal Stress Based on 2050 Changes in the THI
percent

Risk class	10th percentile	90th percentile
Low	<3	<3
Moderate	<3	Between 3 and 4
High	Between 3 and 4	Between 3 and 4
Very high	Between 3 and 4	>4
Extremely high	>4	>4

Note: THI = temperature-humidity index. The values in the table refer to changes of THI from the historical average. Threshold values of 3 and 4 percent were chosen after analyzing the value distributions of percent changes. Values refer to changes in 2050: in the short term it appears that the whole of Nigeria will suffer from a slight THI increase.

Because they are highly dependent on temperature, the results show negative changes in GPP, except for the driest season, when GPP rises, especially in the north, even if there the percentages are highly influenced by very low absolute GPP values. Thus, the driest season is excluded from these analyses.

To capture the variability of GPP across climate models, classes of risk were defined (table 5.3) as a function of agreement between climate models on

Map 5.3 Distribution of Classes of THI Increase in 2050 Compared to 1990

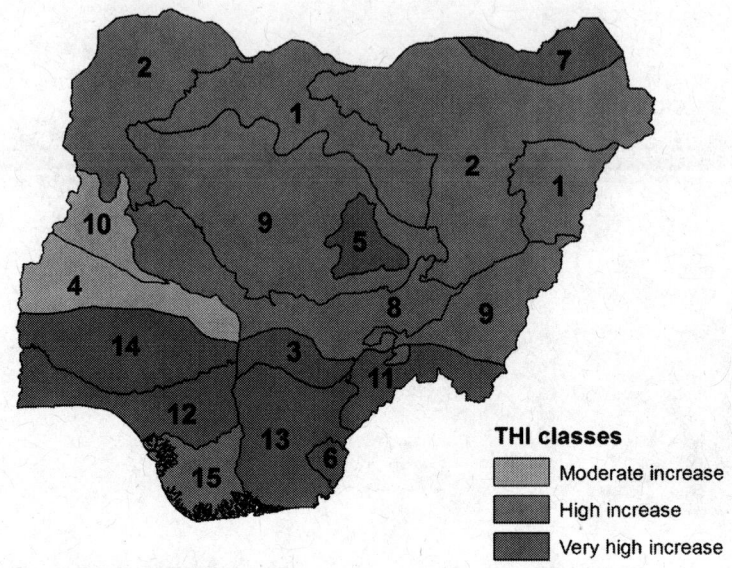

THI classes
- Moderate increase
- High increase
- Very high increase

Source: Authors' calculations based on data sources listed in table 3A.1.
Note: Numbered units are agro-ecological subzones (AESZs). THI = temperature-humidity index.

the magnitude of change over time. The consensus band was defined by the 10th and 90th percentile in the distribution of climate models.

The results are reported in map 5.4 for 2050. Subject to the caveats summarized in box 5.2, uncertain conditions or moderate GPP decreases appear to prevail in the north, mirroring a similar pattern for rainfall. On the other hand, climate models appear to agree on projecting high or very high GPP decreases in the central belt. The southwest seems less likely to experience significant GPP decreases.

Qualitatively combining suitability and sustainability classifications, classes of an integrated risk index were formulated (map 5.5). The index showed that the southern part of the central belt seems to be moving toward high to extremely high unsafe conditions for livestock.

Water Resources

The hydrological modeling tools described in chapter 3 were used to convert changes in climate variables (temperature, precipitation) into changes in water flows, and thus changes in water potentially available for storage or other uses.

Using the same risk classes defined for the analysis of rainfall changes (chapter 4) · to summarize the consensus among climate models, it was found (map 5.6) that by 2020 conditions in 62 percent of the country are expected to be wetter, in 4 percent drier, and in 23 percent stable. In the remaining 11 percent projections are uncertain. In 2050 a significant part of the country

Box 5.1 Thermal Stress and Effects on Livestock

Two studies focusing on Mediterranean areas (Segnalini *et al.* 2011; Vitali *et al.* 2009) demonstrated that daily temperature-humidity index (THI) values are highly correlated to dairy cow mortality rates. Based on observations for 2002–07, they found that THIs of 79.6 and 70.3 represented the break points where there is a significant change in the slopes describing the relationship between maximum or minimum THI and the number of dairy cows that die.

A study of dairy cows in Turkey (Akyuz, Boyaci, and Cayli 2010) demonstrated that milk production is not affected for THI below 72 but both milk production and feed intake start declining above 72 and the decline accelerates above 76. In South Africa, Du Preez, Giesecke, and Hattingh (1990) showed that milk yield declined 10–40 percent in summer compared to winter.

Other studies have established maximum THIs for other productive or physiological functions in dairy cows (milk yield and reproduction) that may be adversely affected by heat stress (Kadzere, Murphy, Silanikove, and Maltz 2002; West, Mullinix, and Bernard 2003). However, Nienaber and Hahn (2007) indicated 79 as the THI danger threshold for the respiration rate of cows. Furthermore, Silanikove (2000) indicated that a THI near 80 would be noxious for most domestic ruminants and >80 would be extremely so.

THI effects are region-specific and are less known for tropical zones, so applying findings from case studies to other areas with different geo-climatic conditions is not warranted. Both animal species themselves and their adaptation to climate conditions can be different and highly influence results. Assigning a risk ranking to selected THI values or intervals would require analyzing sufficiently long time series of observations (e.g., at least 10 years) so as to evaluate relationships among livestock health, mortality, productivity, and daily temperature-humidity conditions. This would in turn require site-specific meteorological variables such as temperature and relative humidity and livestock census information.

Table 5.3 Risk Classes of Impacts on Feed Availability in Relation to 2050 GPP
percent

Risk class	10th percentile	90th percentile
Uncertain	< –45	> –15
Low	> –15	> –15
Moderate	Between –45 and –15	> –15
High	Between –45 and –15	Between –45 and –15
Very high	< –45	Between –45 and –15
Extremely high	< –45	< –45

Note: GPP = gross primary productivity. The values in the table refer to changes of the temperature-humidity index from the historical average. The threshold values of minus 15 percent and minus 45 percent were chosen after analyzing the value distribution of percent changes.

Map 5.4 Distribution of Classes of GPP Decrease in 2050 Compared to 1990

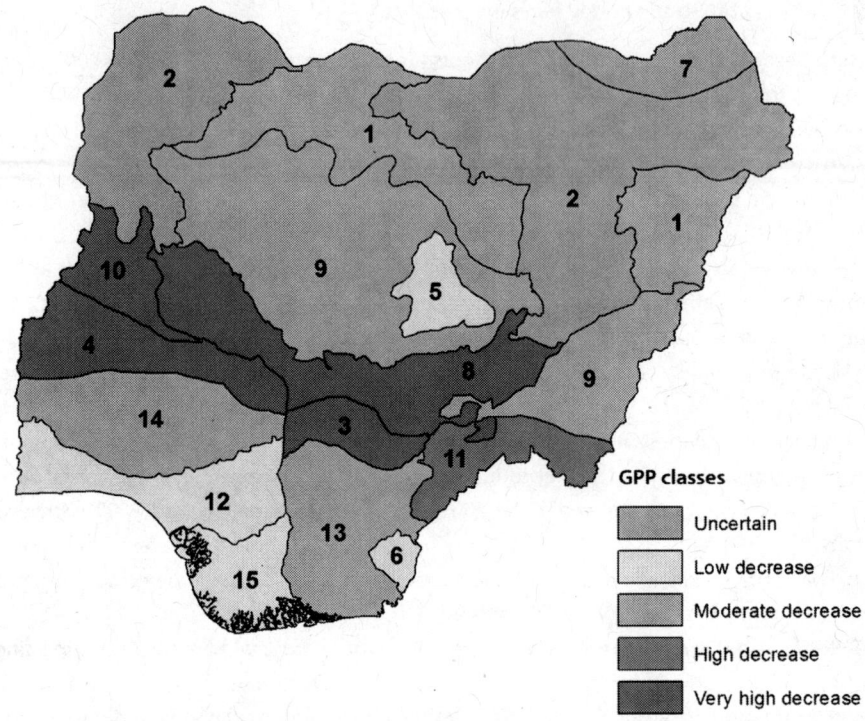

Source: Authors' calculations based on data sources listed in table 3A.1.
Note: GPP = gross primary productivity. Numbered units are agro-ecological subzones (AESZs).

Box 5.2 Gross Primary Productivity and CO_2 Concentration

The analysis of GPP raises the question of whether rising atmospheric CO_2 concentrations might eventually begin to impact carbon-to-nitrogen ratios in forage and hence alter the digestibility and utilization efficiency of forage, which would undermine ruminant productivity. In general, Nigerian grassland is composed mainly of C4 species; C3 species are scarce. C4 and C3 species differ in carbon metabolism and responses to CO_2 levels. In particular C4 grasses are less sensitive to CO_2 exposure and are not likely to change the carbon/nitrogen ratio in the biomass.

Since the specific composition of grass species in Nigeria is not known in detail and the CO_2 fertilization effect is still not fully understood, the analysis did not take this effect into account. However, for livestock systems, adaptation responses may include higher stocking rates because of less intake of lower-quality forage, and dietary supplementation may be used to maintain current production levels, though that will raise the cost of production. Silvopastoral systems can enhance carbon sequestration, and provide both high-nitrogen forage and shade to minimize heat stress.

Map 5.5 Distribution of Integrated Risk for Livestock in 2050 Compared to 1990

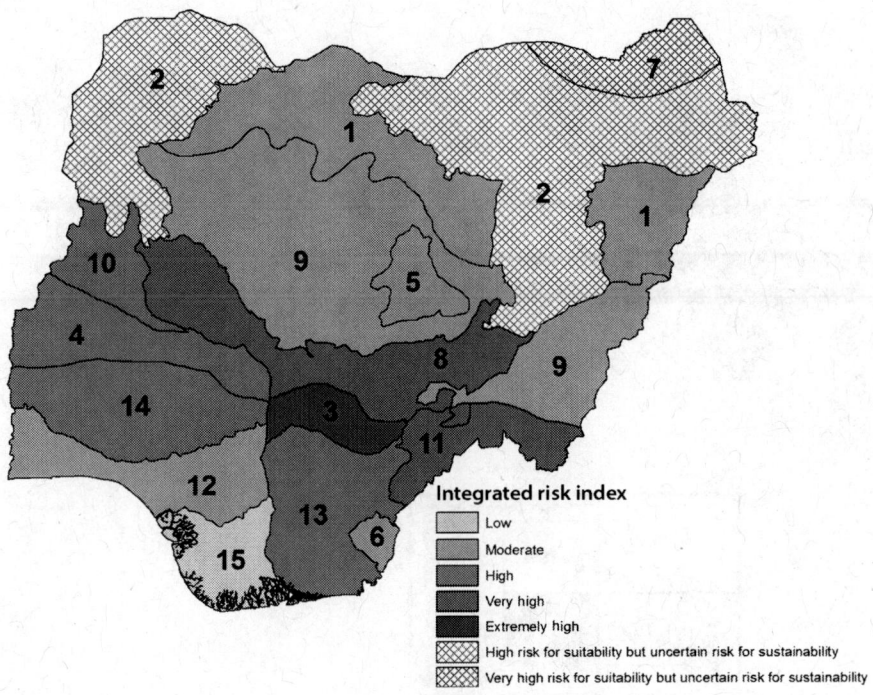

Source: Authors' calculations based on data sources listed in table 3A.1.
Note: Numbered units are agro-ecological subzones (AESZs).

is still projected to become wetter (though decreasing from 62 to 49 percent of land area); and the share of areas at risk of drying increases from 4 to 10 percent (accounting, however, for 17 percent of historical runoff). The share of stable sub-basins drops to 8 percent of total land area and uncertainty increases considerably, to 33 percent of total area. Particularly noteworthy is the high uncertainty for the arid/hyper-arid regions (HA8) in the northeast.

Except for the central high plateau most of central and northern Nigeria shows water resources to be increasingly available, although the uncertainty for 2050 is pronounced. The results for the central area, southeast mountains, and southwest littoral show a general drying trend for 1990–2020 and 1990–2050.

The finding that in 2050 water flows are expected to increase in about half the country is relatively robust to the selection of the stability band (figure 5.4). While the percentage decreases to about 35 percent for a stability band of ±10 percent, it fluctuates only slightly around 50 percent for bands of ±15, 20, and 25 percent. The share of land subject to uncertainty decreases with the wider stability bands; and the share of land exposed to "dry" risks varies between 10 and 2 percent (3–17 percent in terms of historical flow), depending on the stability band selected.

Map 5.6 Distribution of Classes of Risk for Water Flows, 2020 and 2050 Compared to 1990

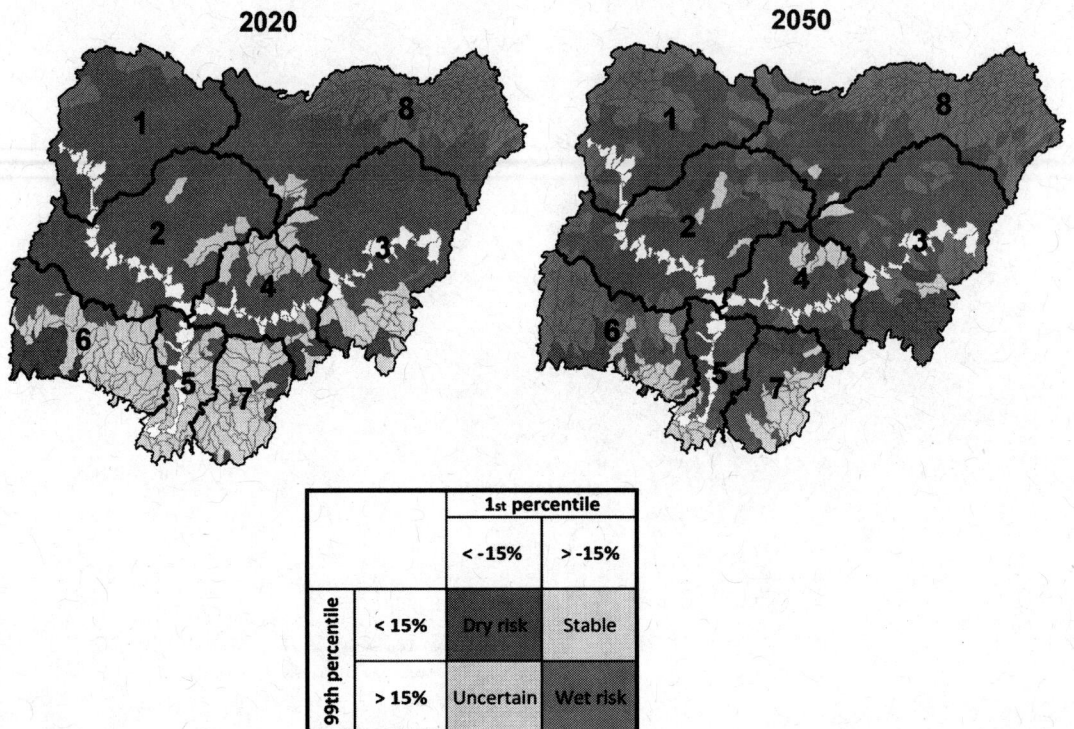

		1st percentile	
		< -15%	> -15%
99th percentile	**< 15%**	Dry risk	Stable
	> 15%	Uncertain	Wet risk

Source: Authors' calculations based on data sources listed in table 3A.1.
Note: Distribution of classes of risk for water resources in 2020 (left) and 2050 (right) compared to 1990. Small units are sub-basins; numbers indicate hydrological areas. White areas are adjacent to the Niger and Benue main river stem and not part of the analyses.

It is important to put into perspective the findings that half of Nigeria may by 2050 expect more water flows. To do that, figure 5.5 plots the largest (99th percentile) projected flow increase against cumulative share in historical flow and relative size of flow in each basin, expressed as a fraction of the maximum flow per unit of land area.

The largest increases in flow are projected to take place in relatively drier basins. While flow is projected to increase up to 200 percent in some such cases, the weighted average increase is only about 33 percent. Only for basins in the bottom 30 percent of the flow distribution is flow projected to increase by more than 30 percent.

Hydropower and Irrigation

The changes in water flows described are likely to have significant effects on the reliability of hydropower and irrigation systems, which are a function of both average magnitude of inflow to a dam and inflow variability. Since these effects are highly site-specific, this section presents the results of the analysis of six hydropower schemes; for irrigation, the analysis of the Tiga scheme that follows

Figure 5.4 Classes of Risk for Water Flows (2050) in Relation to the Stability Band

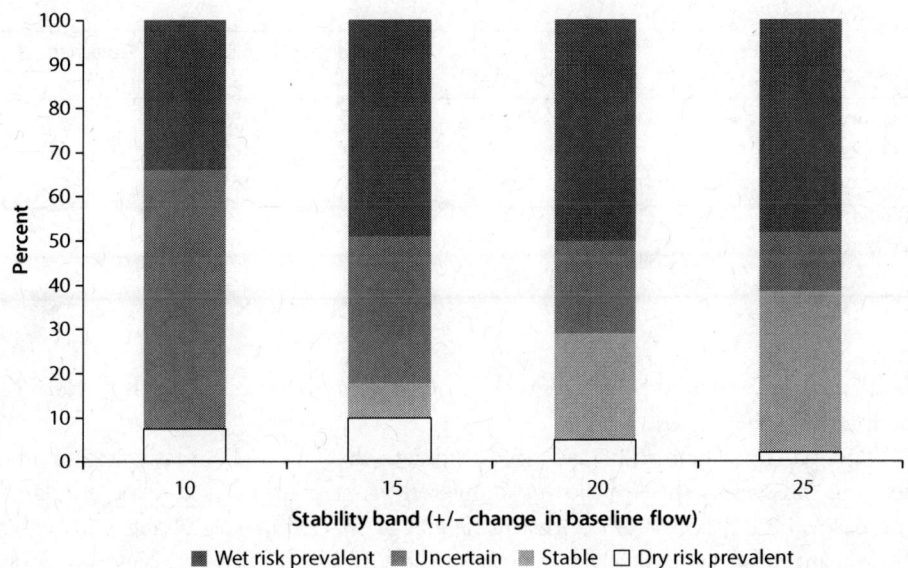

Source: Authors' calculations based on data sources listed in table 3A.1.

Figure 5.5 Distribution of Largest Increase in Water Flow, 2050

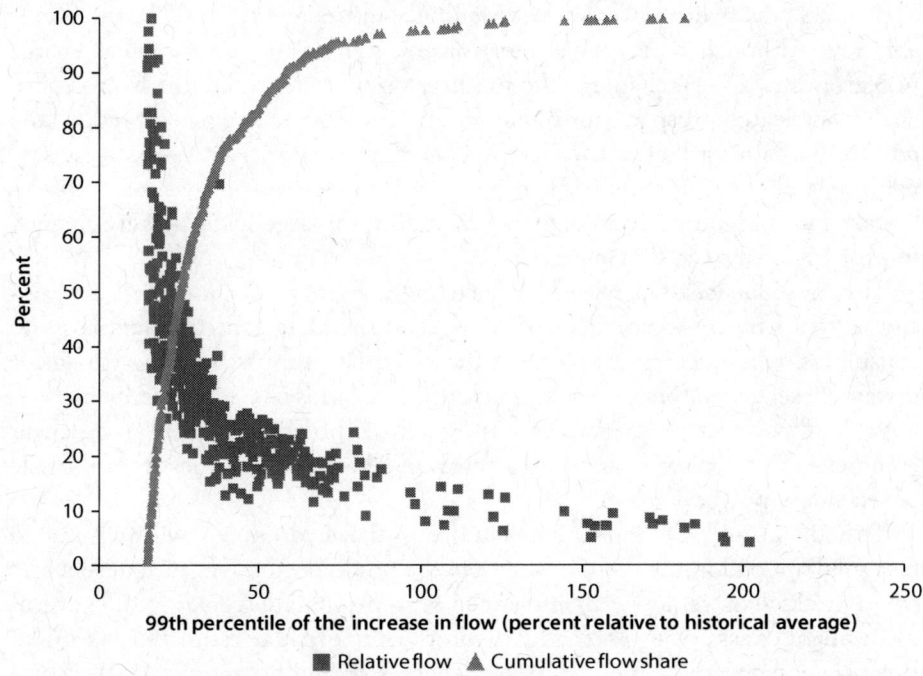

Source: Authors' calculations based on data sources listed in table 3A.1.

Table 5.4 Changes in Reliability for the Full Range of Climate Projections

| | | | | | Reliability | | |
Site	Type	Firm power (GWh/year)	Baseline	2020	Projections of less reliability (%)	2050	Projections of less reliability (%)
Ikere Gorge	Multipurpose	25.1	0.98	0.99–1.00	0	0.99–1.00	0
Shiroro	Hydropower	1,573	0.93	0.68–1.00	9	0.89–1.00	18
Zungeru	Hydropower	2,006	0.92	0.86–1.00	9	0.65–1.00	18
Gurara	Multipurpose	86.3	0.97	0.95–1.00	27	0.94–1.00	27
Dadinkowa	Multipurpose	186	1.00	1.00–1.00	0	0.99–1.00	9
Mambilla	Hydropower	4,369	0.97	0.86–1.00	36	0.77–1.00	45

Source: Authors' calculations based on data sources listed in table 3A.1.

will be complemented by a larger selection of sites evaluated through the lens of adaptation (chapter 6).

The hydropower results (table 5.4) indicate that in most climate change projections, the reliability of power delivered from current and planned dams increases. This is consistent with the results of the analysis of changes in water flows, since the schemes analyzed are in areas where climate change models indicate mainly wetter conditions.

However, in some climate models power delivery becomes less reliable and is sometimes below acceptable safety levels. And risks appear to increase from the medium (2020) to the longer term (2050), in terms of both the number of models indicating less reliability and the severity of worst-case outcomes (figure 5.6). In the most extreme climate change scenario for 2050, the reliability of planned new hydropower sites at Mambilla and Zungeru would be unacceptably low. Although manageable, the risks are significant for Mambilla, a large (4.3 gigawatt [GW]) scheme in the southeast mountains, where the hydrological analysis indicated prospects for stable or drier conditions. That would reduce the reliability in almost half of the climate change projections, with worst-case scenario reliability being reduced by more than 20 percent. It is essential that this be taken into account in the preparation studies for large hydropower schemes, to properly analyze and manage risks.

The consequence of increased failure to deliver firm power, if Nigeria continually suffers from power deficits, is more load-shedding that transfers costs to consumers, who need to supplement the electricity supply by using expensive private diesel generators. However, in climate scenarios characterized by increasing runoff, as most projections indicate, there is potential not only to deliver commercial hydropower but also to generate additional revenues from the sale of secondary power.

The conclusion from the schemes studied is that hydropower will continue to be a solid option for low-carbon growth in the power sector. The current schemes could produce more power than planned in some climate scenarios; the government might consider developing full hydropower potential, estimated at 12 GW. But planning and realization of future schemes should be optimized to accommodate the risk of decreased and more variable inflows, to avoid the chance that

Figure 5.6 Risk of Less Reliable Power Delivery in Six Schemes, 2020 and 2050

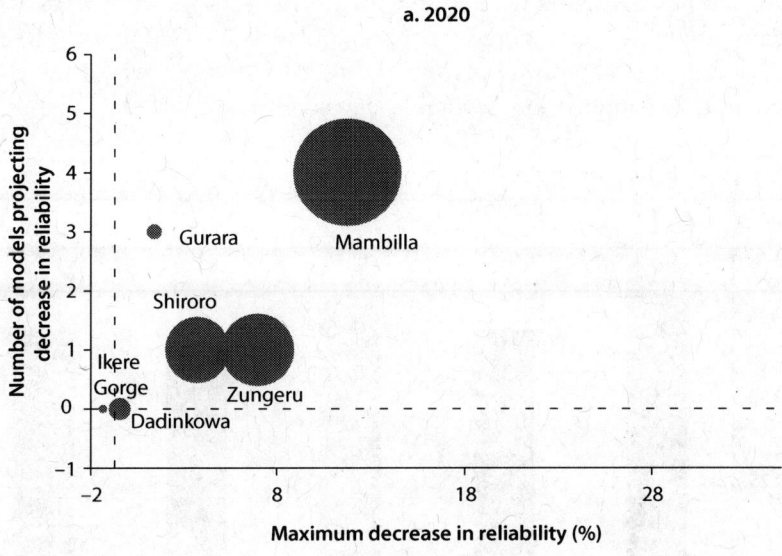

a. 2020

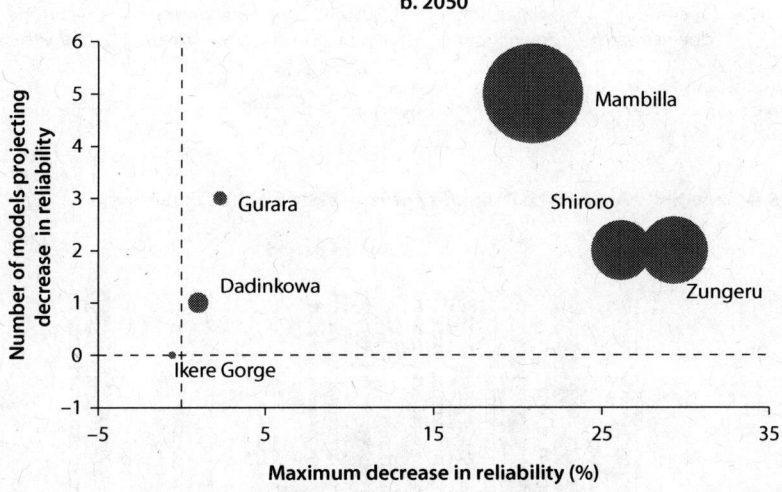

b. 2050

Source: Authors' calculations based on data sources listed in table 3A.1.
Note: The size of the bubbles is proportional to the amount of generation capacity.

reliability drops to unacceptable levels. Care should be taken in the hydrological analysis of these schemes; and in the design phase, historical records should be augmented by information on climate change projections.

The analysis of the single-purpose irrigation scheme at Tiga shows that in terms of historical records the firm yield (80 percent) for an irrigation area of 15,000 hectares (ha) will be acceptable if current water supply and downstream requirements are met. The plans to extend this to 22,000 ha or to manage the planned expansion of downstream water demand might, however, be questionable if the historic climatic conditions prevail (see figure 5.7).

The Tiga irrigation scheme is located in the Kano River, which flows into Lake Chad, in the northern area where the climate change assessment is that future conditions will be wetter (figure 5.8). The simulation for Tiga thus shows a general increase in reliability to supply the 15,000 ha. However, even if the climatic conditions are becoming more favorable, development of further irrigation areas

Figure 5.7 Historic Reliability of the Tiga Scheme, 1976–2006

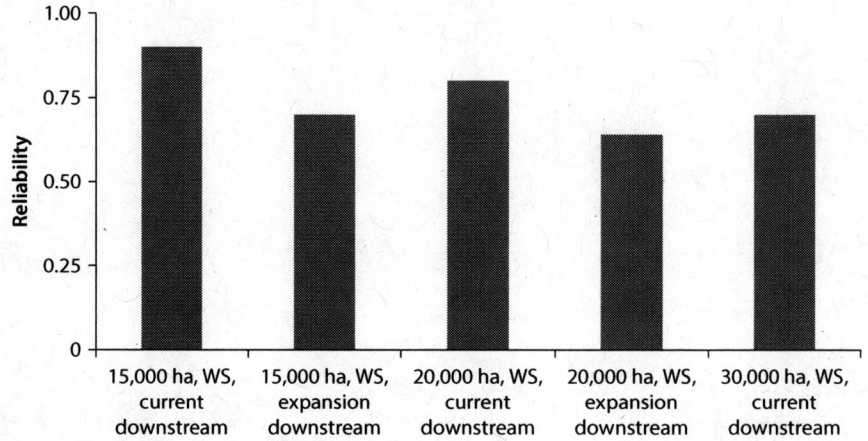

Source: Authors' calculations based on data sources listed in table 3A.1.
Note: Historic reliability of the Tiga scheme (1976–2006) is illustrated for different assumptions about irrigation areas and water supply (WS).

Figure 5.8 Sensitivity Analysis of Reliability of Firm Yield for the Tiga Scheme

a. 2006–35 period

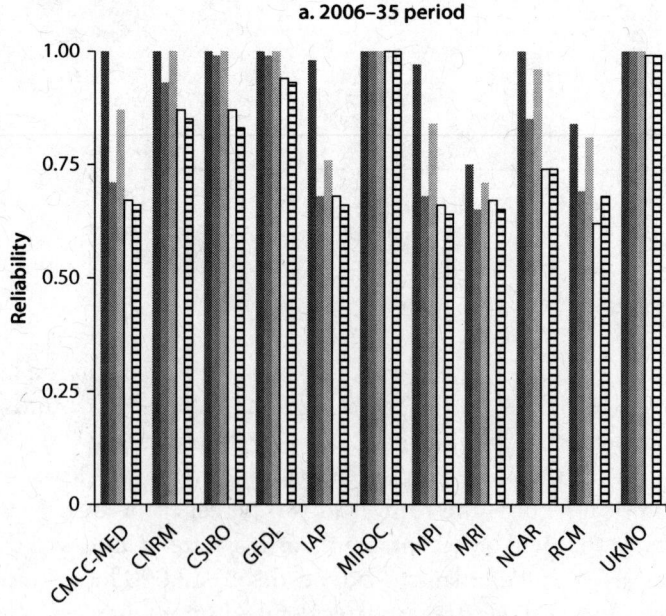

figure continues next page

Figure 5.8 Sensitivity Analysis of Reliability of Firm Yield for the Tiga Scheme *(continued)*

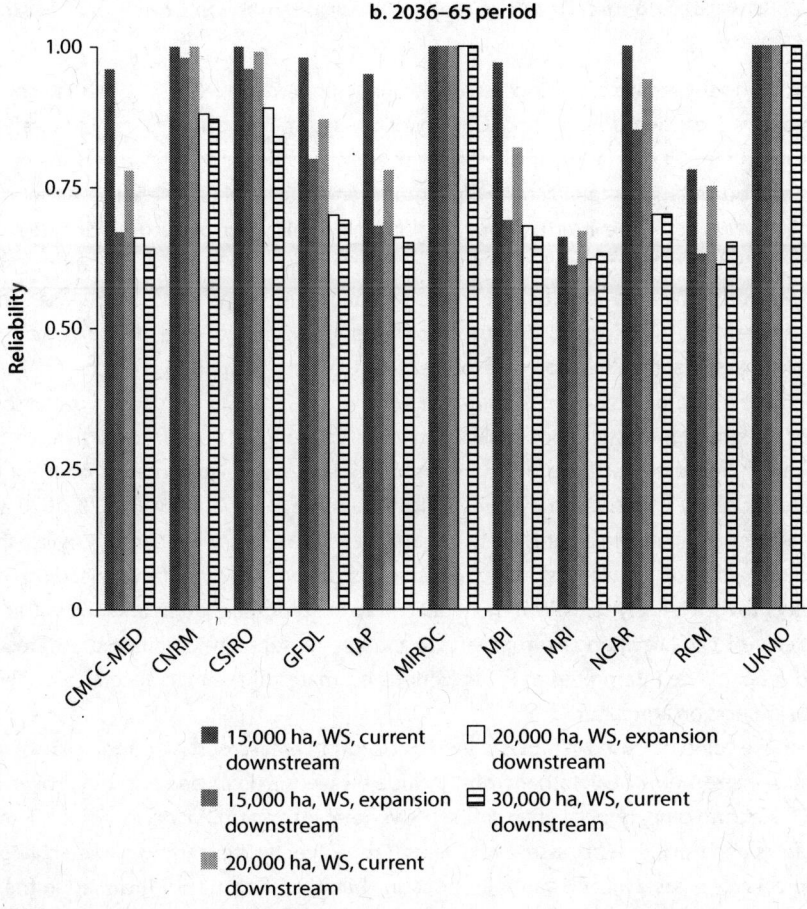

b. 2036–65 period

Legend:
- 15,000 ha, WS, current downstream
- 15,000 ha, WS, expansion downstream
- 20,000 ha, WS, current downstream
- 20,000 ha, WS, expansion downstream
- 30,000 ha, WS, current downstream

Source: Authors' calculations based on data sources listed in table 3A.1.
Note: RCM = Regional Climate Model; CMCC-MED = Euro-Mediterranean Center on Climate Change; CNRM = Centre National de Recherches Météorologiques; CSIRO = Commonwealth Scientific and Industrial Research Organization; GFDL = Geophysical Fluid Dynamics Laboratory; IAP = Institute of Atmospheric Physics; MIROC = Center for Climate System Research; MPI = Max Planck Institute; MRI = Meteorological Research Institute; NCAR = National Center for Atmospheric Research; UKMO = United Kingdom Meteorological Office; WS = water supply.

or significantly expanding downstream water demand is questionable because in only a minority of the climate change projections is such an expansion acceptably reliable. As with the hydropower schemes, care must be taken when planning and putting in place future irrigation infrastructure; future climate risks need to be assessed.

Macroeconomic Impacts

This section presents results obtained by applying to the macroeconomic model described in chapter 3 the changes in crop yields discussed above (see box 5.3). To bracket climate model uncertainty, taking into account the time involved in

Box 5.3 Interpreting Results of the Macroeconomic Analysis: Conceptual Issues and Terminology

In interpreting the results of computable general equilibrium models (CGEs) like the one used in this analysis, caution is warranted. CGEs typically assume that markets function relatively smoothly and that it is easy to adjust external shocks by reallocating resources across sectors, inputs, and outputs, so as to maximize firms' profits and households' well-being. Although the CGE used in this book, the Intertemporal Computable Equilibrium System (ICES) model, has a number of features intended to better approximate the reality of Nigeria's agriculture (including, for example, the agro-climatic constraints to reallocating crops across space—see appendix I for details), it is difficult to fully capture the range of market imperfections that might preclude the efficient reallocation of resources that CGE models tend to project.

The model also assumes—in accordance with Vision 20: 2020—a decline over the long term in the gross domestic product (GDP) share of agriculture from 40 to 15 percent and a significant increase in baseline productivity, resulting largely from expanding irrigation.

If the economy diversifies away from agriculture more slowly than Vision 20: 2020 anticipates, a larger agriculture sector is likely to transmit to the rest of the economy larger shocks from climate change. Similarly, slower expansion of irrigation (or other factors leading to productivity growth) is likely to make it more difficult to meet domestic demand and will amplify the effects of climate change on imports. For these reasons, the findings presented here should probably be interpreted as lower-bounds estimates of the macroeconomic effects of climate impacts on agriculture.

On the terminology used in tables 5.5–5.7: the group "cereal crops" is an aggregate of millet, sorghum, maize, and wheat; "other crops" include oil seeds, sugar cane and sugar beet, land-based fibers, and other nonclassified crops. The vegetables and fruits category incorporates all vegetables and fruits except cassava and yams. Crops directly affected by climate change are rice, cereal crops, cassava, and yams. Production changes for crops not included in the yield modeling were driven by market reactions to relative price changes endogenously determined by the economic model.

---

running the economic model multiple times, the effects were evaluated with respect to two scenarios of climate effects on yields, one more pessimistic (high impact) and the other more optimistic (low impact). Unless otherwise noted, all results—which assume CO_2 constant concentration (no fertilization effects, see box 5.2)—are expressed as percent deviations of the variables from the value estimated in the no-climate change reference scenario.

Subject to the caveat in box 5.3, the model projects the following responses to climate shocks on crop yields:

- A decline in production, growing over time and particularly significant by 2050, of the "cereal crops" aggregate, which unlike the other categories is about minus 9.6 percent even in the more optimistic climate scenario (table 5.5). Low-case scenario declines are also high for rice (minus 5.9 percent).

Table 5.5 Change in Production Volume by Crop, Year, and Climate Scenario
percent

	2020		2050	
Crop	Higher impacts	Lower impacts	Higher impacts	Lower impacts
Cassava	−0.3	1.0	−4.8	−3.0
Cereal crops	−2.4	−1.6	−15.8	−9.6
Other crops	−0.6	1.5	−6.6	−4.7
Rice	−1.0	0.2	−8.2	−5.9
Vegetables and fruits	−0.3	0.6	−4.1	−2.8
Yams	−0.2	0.9	−4.7	−3.1

Source: Authors' calculations based on data sources listed in table 3A.1.
Note: For the definition of crops, see the last paragraph of box 5.3.

Table 5.6 Change in Prices by Crop, Year, and Climate Scenario
percent

	2020		2050	
Crop	Higher impacts	Lower impacts	Higher impacts	Lower impacts
Cassava	0.90	−10.40	14.60	1.80
Cereal crops	2.50	3.10	14.50	7.70
Other crops	−0.40	−0.60	−1.10	−0.50
Rice	10.20	8.10	21.88	16.03
Vegetables and fruits	−0.20	−0.90	−2.30	−1.30
Yams	−0.80	−7.20	10.00	4.60

Source: Authors' calculations based on data sources listed in table 3A.1.
Note: For the definition of crops, see the last paragraph of box 5.3.

Table 5.7 Change in Net Imports by Crop, Year, and Climate Scenario
percent

	2020		2050	
Crop	Higher impacts	Lower impacts	Higher impacts	Lower impacts
Cassava	0.90	−17.60	19.20	−1.70
Cereal crops	2.40	4.00	12.30	5.80
Other crops	−1.30	−0.20	−8.70	−5.60
Rice	10.00	9.10	43.25	37.83
Vegetables and fruits	−1.20	−1.10	−12.50	−8.10
Yams	2.80	30.10	−26.80	−13.60

Source: Authors' calculations based on data sources listed in table 3A.1.
Note: For the definition of crops, see the last paragraph of box 5.3.

- An increase in domestic crop prices (table 5.6), which will be particularly severe for rice, suggesting more rigid demand.
- Significant changes in food trade patterns (table 5.7), with net imports of rice and the cereal crops aggregate rising to offset the projected decline in domestic production as the population, and thus food demand, grow.

Figure 5.9 Deviation of GDP from the No Climate Change Reference Scenario

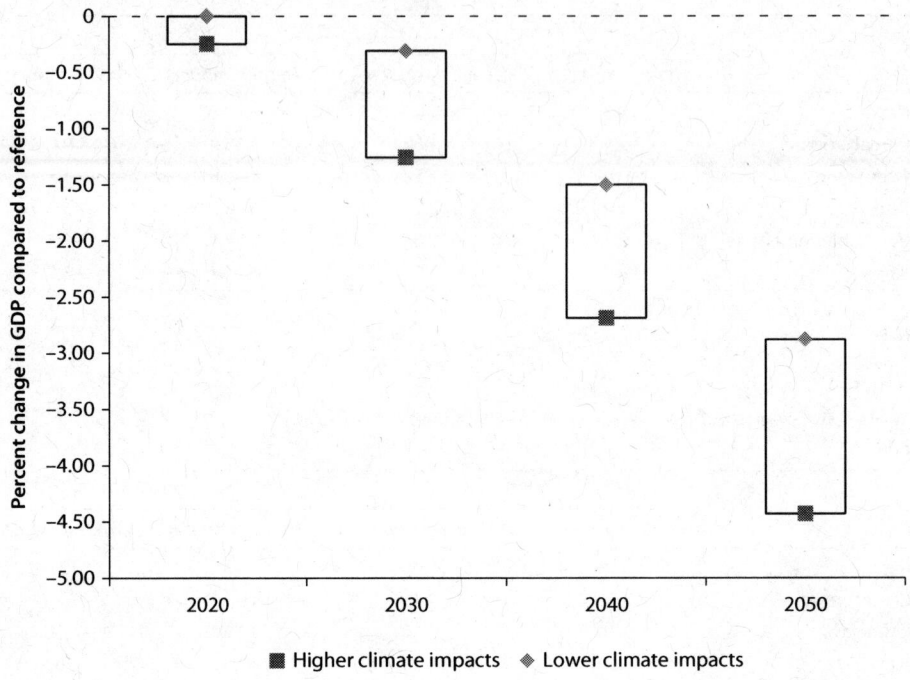

Source: Authors' calculations based on data sources listed in table 3A.1.
Note: The figures indicate the range of impacts for the end-year of each period.

Rice and cereals represent by far the majority of agricultural imports in Nigeria in the baseline (35 percent rice and 46 percent cereals in 2050). Accordingly, the general equilibrium adjustment to the decline in production occurring for all crops in 2050 consists in meeting demand where possible by increasing imports, demand being higher for crops with relatively lower import prices in the baseline, such as rice and other cereals.

The combined effect of changes in production, prices, and imports produces a reduction by 2050 in total GDP compared to the no climate change reference scenario of 3–4.5 percent (figure 5.9), depending on the climate model. For the reasons set out in box 5.3, these figures should probably be considered conservative, lower-bound estimates of the macroeconomic impacts of climate change.

Note

1. As described in chapter 3, MAR measures whether the population can fulfill its dietary requirements for energy and nutrients from the available food crop. MAR is calculated by averaging individual nutrient adequacy ratios (NAR; Hatløy, Torheim, and Oshaug 1998). NAR specific to calories or nutrients is defined as the ratio of energy or nutrients available per person from food crop quantities to the respective recommended nutrient intakes (RNI). For details, see appendix D.

References

Akyuz, A., S. Boyaci, and A. Cayli. 2010. "Determination of Critical Periods for Dairy Cows Using Temperature Humidity Index." *Journal of Animal and Veterinary Advances* 9 (13): 1824–27.

Du Preez, J. H., W. H. Giesecke, and P. J. Hattingh. 1990. "Heat Stress in Dairy Cattle and Other Livestock under Southern African Conditions. I. Temperature Humidity Index Mean Values During the Four Main Seasons." *Onderstepoort Journal of Veterinary Research* 57: 243–48.

Hatløy, A., L. E. Torheim, and A. Oshaug. 1998. "Food Variety—A Good Indicator of Nutritional Adequacy of the Diet? A Case Study from an Urban Area in Mali, West Africa." *European Journal of Clinical Nutrition* 52: 891–98.

Jones, P. G., and P. K. Thornton. 2003. "The Potential Impacts of Climate Change in Tropical Agriculture: The Case of Maize in Africa and Latin America in 2055." *Global Environmental Change* 13: 51–59.

Kadzere, C. T., M. R. Murphy, N. Silanikove, and E. Maltz. 2002. "Heat Stress in Lactating Dairy Cows: A Review." *Livestock Production Science* 77: 59–91.

Lobell, D. B., and M. B. Burke. 2010. "On the Use of Statistical Models to Predict Crop Yield Responses to Climate Change." *Agricultural Forest Meteorology* 150: 1443–52.

Lobell, D. B., M. B. Burke, C. Tebaldi, M. D. Mastrandrea, W. P. Falcon, and R. L. Naylor. 2008. "Prioritizing Climate Change Adaptation Needs for Food Security in 2030." *Science* 319: 607–10.

Nienaber, J. A., and G. L. Hahn. 2007. "Livestock Production System Management Responses to Thermal Challenges." *International Journal of Biometeorology* 52: 149–57.

Segnalini, M., A. Nardone, U. Bernabucci, A. Vitali, B. Ronchi, and N. Lacetera. 2011. "Dynamics of the Temperature-Humidity Index in the Mediterranean Basin." *International Journal of Biometeorology* 55: 253–63. doi:10.1007/s00484-010-0331-3.

Silanikove, N. 2000. "Effects of Heat Stress on the Welfare of Extensively Managed Domestic Ruminants." *Livestock Production Science* 67: 1–18.

Thornton, P. K., P. G. Jones, G. Alagarswamy, and J. Andresen. 2009. "Spatial Variation of Crop Yield Response to Climate Change in East Africa." *Global Environmental Change* 19 (1): 54–65.

Vitali, A., M. Segnalini, L. Bertocchi, U. Bernabucci, A. Nardone, and N. Lacetera. 2009. "Seasonal Pattern of Mortality and Relationships between Mortality and Temperature Humidity Index in Dairy Cows." *Journal of Dairy Science* 92: 3781–90.

West, J. W., B. G. Mullinix, and J. K. Bernard. 2003. "Effects of Hot, Humid Weather on Milk Temperature, Dry Matter Intake, and Milk Yield of Lactating Dairy Cows." *Journal of Dairy Science* 86: 232–42.

CHAPTER 6

Adaptation Options in the Agriculture and Water Sectors

This chapter identifies options to reduce the impacts of climate change already discussed on agriculture, water resources, and economic growth and evaluates them in technical terms; the institutional challenges to be addressed to enable their adoption are discussed in chapter 7.

Adaptation in Agriculture

In general terms it seems plausible that a sound adaptation strategy will include a combination of expansion of irrigated areas and improved management of rain-fed crops. A major policy question, then, is the right mix of these two approaches. Several factors, such as relative costs, resource availability, and the institutional context, will help in determining the ultimate outcome, with sector policies likely to play an important but not unique role. It is nevertheless informative, at the level of aggregation presented here, to explore illustrative options, subject to the data and the modeling tools the study was able to mobilize given the time available.

If Nigeria in the medium to long term achieves a target of irrigating 25 percent of total crop land (some 11 million hectares [ha]), it is plausible that most adaptation will take place through enhancing the resilience of the 75 percent of areas still under rain-fed cultivation. This section therefore analyzes options that can be deployed in rain-fed areas, the extent to which they could counter the impact of climate change on production, and at what cost. A full offset of climate impact might require, in addition to targeted efforts in rain-fed areas, further expansion of irrigation. That could be economically advantageous as long as the unit costs of various options can be held in check.

Sustainable Land Management Practices

Employing a range of farming practices (box 6.1) consistent with principles of sustainable land management (SLM) is likely to deliver both higher yields and enhanced resilience to climate variability.

Box 6.1 Sustainable Land and Water Management Practices

Conservation agriculture: This term refers to a basket of technologies such as those described below, as well as crop rotation and no tillage. No tillage has a cost-benefit ratio much higher than 1, hence its recent widespread adoption in Brazil, Mexico, and the southeast United States. It is generally appropriate for rich soils where it can boost yields by up to 60–80 percent while storing soil carbon (itself an adaptive metric since carbon boosts soil moisture retention). The technique has not been carried out at scale in Nigeria and will likely require trials in various settings to determine not only the agronomic parameters but also the socioeconomic preferences of farmers, as there may be implications related to labor and use of chemicals to reduce pests and weeds—not to mention how much appetite farmers have to change their entire farm production system. No tillage, and conservation agriculture broadly, will require a major and well-funded extension effort (see below). A transition to conservation agriculture has implications for agricultural mechanization. There is a need for implements adapted to the different ecological zones, such as those that disturb the soil only minimally, those for seeding and planting, and those for managing crop residues.

Water harvesting: Low-cost and small-scale on-farm water harvesting, especially in the north and central belt, can serve as both an intermediate technology in the absence of larger-scale irrigation and to increase the efficiency of water use on irrigated farmland. For example, planting pits surrounded by demi-lunes act as tiny catchments to direct and hold water in the soil. Some farmers have added termites to these planting pits to increase soil fertility.

Integrated soil fertility management: Combining organic and inorganic fertilizers in strategic amounts, based on certain combinations of crops and agro-ecological zones (AEZs) and coupled with soil and water-conserving mulch, can help farmers adapt to climate variability while also contributing to soil carbon, soil health, and higher yields. Organic fertilization can come from manure, mulch, crop residues, or nitrogen-fixing trees or legumes. Supported by extension, the benefits of this technology can outweigh the costs in the short, medium, and long terms. It can also reduce the financial burden of private and public expenditure on inorganic fertilizer.

Agroforestry and revegetation: This area covers natural regeneration of tree cover and other agroforestry strategies, such as live fencing, shelterbelts, and woodlots. The adoption of agroforestry methodologies that maintain shade canopies in cocoa plantations can both increase productivity and help mitigate climate change by reducing expansion of cultivation to natural forests. Maintaining shade canopy over cocoa farms is an evolving option for diversifying farm income as well as sequestering carbon. Northern Nigeria could benefit as southern Niger has from encouraging farmer-managed natural generation, which has led to a regreening of Sahelian systems and brought land back into crop, livestock, and firewood production. Other efforts to introduce or reintroduce trees on farms in Nigeria's other AEZs could generate positive benefits in the medium and long terms. The southern multistory farm system common in Igbo communities serves as a buffer to climate risks by diversifying the farmscape. The cost of such agroforestry strategies can vary greatly, from US\$166 a hectare to as much as US\$906,

box continues next page

Box 6.1 Sustainable Land and Water Management Practices *(continued)*

but depending on the technology and local circumstances the benefit-cost ratio can turn positive after four years from adoption.

Restoration of degraded pasture: Important for the north and central belts, this approach requires a combination of technologies (seeding, irrigation) and approaches (laws to address open access, reduce burning, and establish grazing reserves, rotational grazing, or grazing corridors). The government and certainly some livestock-dependent communities are practicing some of these, but not as a strategic package to both reduce climate impact and boost livestock and land productivity. Given the projected impact of climate change on the health and productivity of livestock (thermal stress) and land (gross primary productivity) in the central and northern zones, it is reasonable to conclude that the benefits of intervention outweigh the costs, which begin at about US$80 per ha for the farmer and rise to US$90 in succeeding years, coupled with about US$35 per ha of public expenditure, which drops quickly the next year. Costs are much cheaper without inorganic fertilizers. In terms of benefits, rotational grazing alone has been modeled to show a 33 percent increase in biomass.

Out of the wide range of practices described in box 6.1, several were analyzed in terms of their potential to offset—across space (different AEZs); time (2020 and 2050); and crops—the negative impacts of climate change on yields. The selection was dictated by data availability and by the suitability of an option for integration into the crop modeling used to evaluate climate impacts. For rain-fed areas, seven adaptation options (table 6.1) were analyzed, clustered as follows:

1. Shift of the sowing/planting date (1 month earlier or later than the traditional calendar)
2. Conservation/organic agriculture practices, including manure and residues management
3. Use of inorganic fertilizers.

For irrigated crops, the analysis focused on yield improvements that could be achieved by modifying planting and sowing dates. To address climate model uncertainty, data from the regional climate model (RCM) and the two extreme perturbations were considered. The concentration of CO_2 in the atmosphere was kept constant.

As for the planting date, for each crop the model was run shifting the period 1 month earlier and 1 month later than the traditional cultivation calendar to evaluate the effects on crop yields. In terms of conservation agriculture, the analysis focused on nutrient management, and evaluated the use of manure and residues to complement current nutrient provision (manure 1 and residues 1) or replace them (manure 2). Residues are dead biomass from the previous harvest, not usable commercially but still rich in nutrients to be released back to the soil.

Table 6.1 Adaptation Options Analyzed

Group	Adaptation option	Description	Benefits	Constraints
Rain-fed areas				
Change in planting/ sowing dates	Plus 1 month Minus 1 month	Shift the sowing/planting date 1 month before and 1 month after the ordinary sowing/ planting date.	It may allow avoiding very hot and/or dry periods. It does not imply cost for farmers and can be immediately put in place, if the results are positive. In some agro-ecological subzones (AESZs) and for some crops (cereals), yields have increased 20–30%, depending on crop and AESZ.	Farmers need extensive training and access to skilled advisory services. Results are highly variable depending on the crop and the cultivar.
Inorganic fertilization	Fertilizer 1 Fertilizer 2	Increase by 30% (fertilizer 1) and by 60% (fertilizer 2) over the ordinary fertilization amount.	Yields increase up to 20–30% for cereals and yams, and up to 40% for cassava.	Relatively high cost of fertilizers; farmers need access to skilled advisory services. There may be an impact on the environment.
Conservation agriculture	Manure 1 Manure 2 Residue	Application of manure (manure 1) or residues from crop production (residue) to complement baseline nutrient management; complete substitution of inorganic fertilization with manure (manure 2).	Yields increase up to 25% for sorghum and millet, up to 35% for rice, and up to 50% for maize and cassava.	Farmers need extensive training and access to skilled advisory services. There may be a relatively high up-front cost for the purchase or application of manure and residues.
Irrigated areas				
Combining shift in growing period and irrigation		Shift the sowing/planting date 1 month before and 1 month after the traditional date, in addition to irrigation practice.	Yields increase for cassava and yams, and there is a positive synergy between irrigation and the shift in growing period.	Farmers need extensive training and access to skilled advisory services.

Finally, additional use of inorganic fertilizers was investigated, at a lower (fertilizer 1) and higher (fertilizer 2) intensity.

Considering the total number of outcomes (number of crops × number of AESZs × the three climate models considered), the adaptation options seem to perform well, both in the short term (2020) and the longer term (2050). Yields improved 70–75 percent of the time; however, there are considerable differences in how the individual options performed (figures 6.1 and 6.2), with residues and other nutrient management options delivering improvements at least 90 percent of the time; change in planting dates did not do as well in rain-fed areas but appears to work much better in irrigated conditions.

The range and reliability of yield improvements vary by adaptation option; residues and manure 1 at worst perform slightly less well than the no-adaptation

Figure 6.1 Safety Ratio of the Adaptation Options, 2020

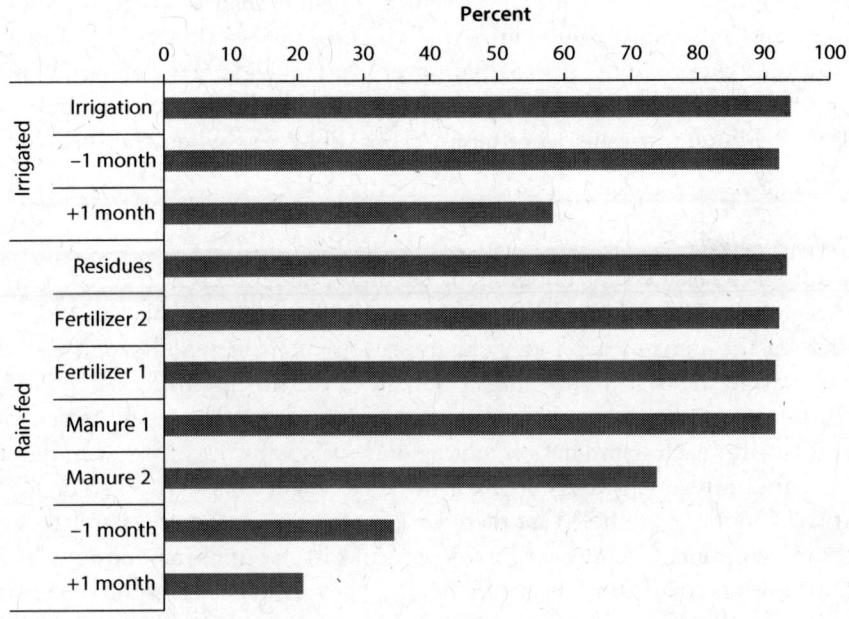

Source: Authors' calculations based on data sources listed in table 3A.1.
Note: The safety ratio is defined as the number of times a given adaptation option increases yield compared to the no-adaptation case, divided by the total number of outcomes.

Figure 6.2 Adaptation Options: Maximum and Minimum Yield Improvement

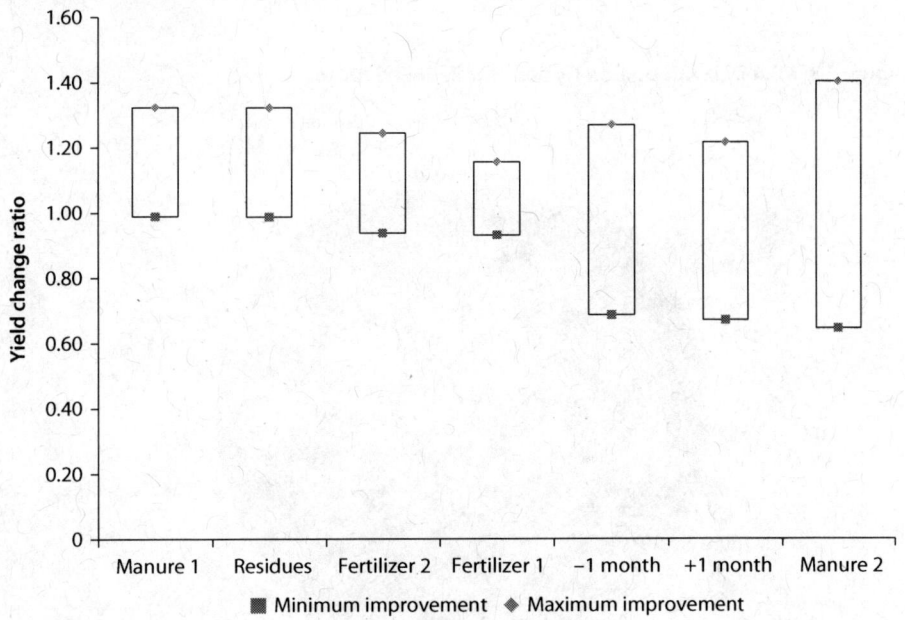

Source: Authors' calculations based on data sources listed in table 3A.1.
Note: The figure reports the ratio between yield change projected with adaptation and yield change with no adaptation. Values above 1 indicate improvements and below 1 worse performance; a value of 1 indicates no improvement.

case; in the best cases, they deliver yields 30 percent higher. Change in planting dates can produce significant improvements (more than 20 percent), but in some crops and zones they can actually cause a yield decline (as much as 30 percent lower than the no-adaptation yield). The wide range of variability in the performance of the options points up the need to further evaluate the suitability of different options to different crops and AEZs when there is climate uncertainty.

Regrets Analysis

The approach selected for undertaking such an evaluation is a "regrets" analysis. The regrets for adopting each option are expressed across the three climate models as the percent gap[1] in yield improvement between the option being examined and the best-performing option; next, the maximum regret was calculated for each option; and finally, the "mini-max" adaptation option was identified—for each combination of crop and AESZ—as the one that minimizes the maximum regrets across climate models (see figure 6.3).

It was found (figure 6.3) that manure 2, manure 1, and residues are the best-performing options, accounting for 75 percent of total mini-max options. These options not only add to nutrient availability, they also increase soil fertility in a broader sense by improving soil physical characteristics and soil water retention and thus availability and by reducing nutrient losses to runoff and leaching.

A key insight from the analysis is that the optimal mix of adaptation option is highly crop- and location-specific. For example (figure 6.4), the mini-max strategy for cassava calls for adopting manure 2 in 90 percent of the AESZ, and

Figure 6.3 Mini-Max Adaptation Options for Rain-Fed Areas

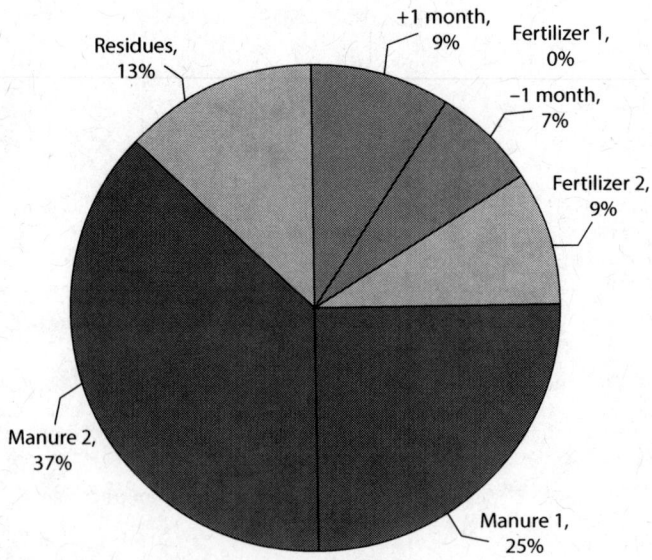

Source: Authors' calculations based on data sources listed in table 3A.1.

Figure 6.4 Composition of Mini-Max Adaptation Strategies across Rain-Fed Crops

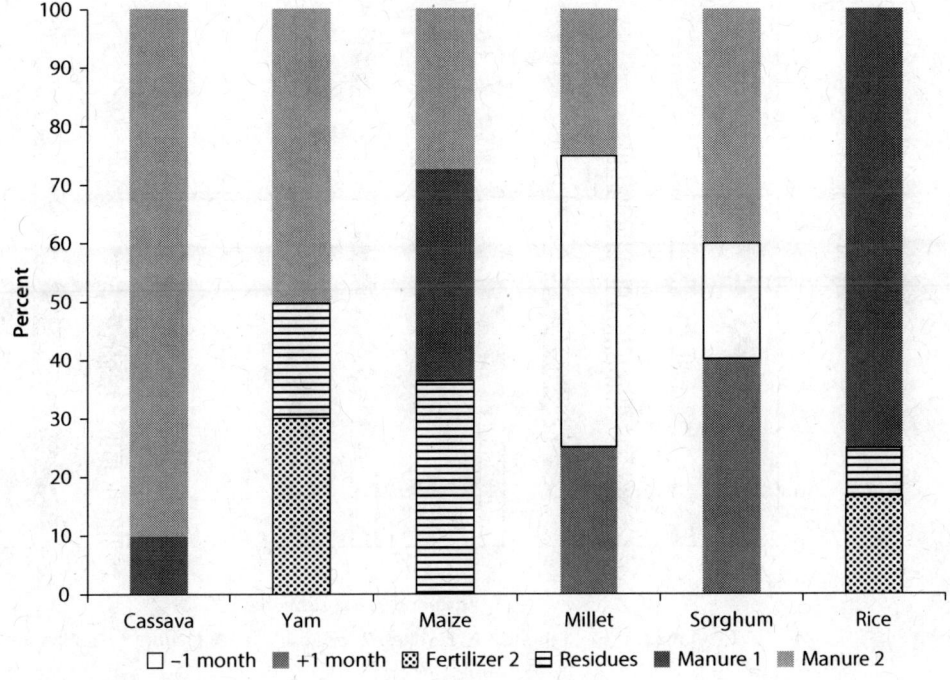

Source: Authors' calculations based on data sources listed in table 3A.1.

manure 1 in 10 percent. For rice, the strategy is to adopt manure 1 in 75 percent of the AESZs, fertilizer 2 in 17 percent, and residues in 8 percent.

Similarly, at the AESZ level (figure 6.5), the mini-max adaptation strategy in AESZ 10 entails adoption of a single option, manure 1, whereas in AESZ 11, the strategy incorporates four options: planting 1 month earlier, fertilizer 2, residues, and manure 2.

These findings highlight the importance of stepping up research, development, and extension services so as to identify and deploy crop- and location-specific adaptation options. More detailed results for the individual adaptation options, separately for each crop and AESZ, are reported in appendix C.

Combining Adaptation in Irrigated and Rain-Fed Areas

Deployed throughout the country, are the adaptation options discussed capable of fully offsetting climate impacts? If not, how much more irrigation might be required? What are the costs associated with a strategy that combines rain-fed adaptation with extra irrigation?

To begin to formulate answers to these questions, this section defines an evolution in 2020 and 2050 of cropping patterns in AEZs using information from the macroeconomic model introduced in chapter 3. It also evaluates the land area to which mini-max adaptation options should be applied (table 6.2) to

Figure 6.5 Composition of Mini-Max Adaptation Strategies across AESZs

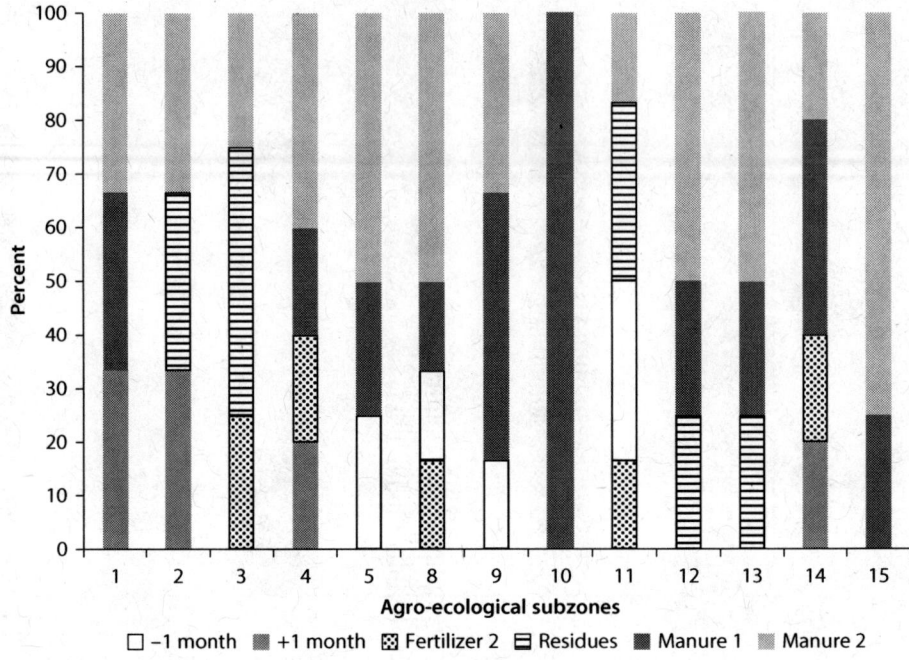

Source: Authors' calculations based on data sources listed in table 3A.1.

Table 6.2 Applying Mini-Max Rain-Fed Adaptation Options by Year and Climate Model
ha, millions

	2020			2050		
Crops	*NCAR*	*GFDL*	*RCM*	*NCAR*	*GFDL*	*RCM*
Cassava	0.00	0.00	0.22	0.00	0.23	2.06
Maize	0.07	0.33	0.18	3.84	4.05	4.05
Millet	0.00	0.00	0.27	3.01	3.16	3.16
Rice	0.17	0.10	0.13	2.29	2.63	2.63
Sorghum	0.36	0.34	0.29	4.01	4.42	4.42
Yams	0.00	0.00	0.02	0.00	1.66	1.66
Total	0.59	0.77	1.11	13.15	16.15	17.98

Source: Authors' calculations based on data sources listed in table 3A.1.
Note: NCAR = National Center for Atmospheric Research; GFDL = Geophysical Fluid Dynamics Laboratory; RCM = Regional Climate Model.

eliminate as much as possible of the production gap between the reference period and the three climate change scenarios selected to define a bracket of climate model uncertainty. Excluding combinations of zones and climate models where an increase in production is expected, the analysis found that by 2020 adaptation should be applied to 0.6–1.1 million ha, depending on the climate model considered; by 2050, due to more severe climate impacts, the area should increase to 13–18 million ha.

Although in 2020, except for millet in one climate model, the mini-max adaptation options fully offset climate impacts, in 2050 there are gaps of 1–22 percent, depending on the crop and the climate model (table 6.3).

Taking into account the yield differential over time between rain-fed and irrigated conditions, the 2050 production gap could be filled before then by expanding irrigation to 1.5–1.7 million ha (table 6.4).

Undertaking adaptation options to the extent indicated by table 6.4 eliminates the production gap created by climate change. But does it make economic sense? The final step of the analysis is to evaluate aggregate costs and benefits of each adaptation strategy. Costs include direct outlays associated with expanding irrigation and promoting improved farming practices in rain-fed areas; to bracket unit cost variability, a range was defined for each cost component, using the sources discussed in box 6.2.

In the analysis, additional irrigation is assumed to come from surface water stored in dams. However, groundwater is already important in irrigation in Nigeria and if properly managed may continue to be a reliable irrigation source.

Table 6.3 Production Gap Eliminated by Mini-Max Rain-Fed Options, by Year and Climate Model

percent

	2020			2050		
Crops	NCAR	GFDL	RCM	NCAR	GFDL	RCM
Cassava	n.a.	n.a.	100	n.a.	100	92.2
Maize	100	100	100	100	99.9	99.1
Millet	n.a.	n.a.	95.1	100	82.6	78.3
Rice	100	100	100	100	89.2	89.0
Sorghum	100	100	100	100	94.0	93.9
Yams	n.a.	n.a.	100	n.a.	97.4	92.3

Source: Authors' calculations based on data sources listed in table 3A.1.
Note: NCAR = National Center for Atmospheric Research; GFDL = Geophysical Fluid Dynamics Laboratory; RCM = Regional Climate Model; n.a. = not applicable.

Table 6.4 Area of Adaptation Application by Climate Model

ha, millions

	2020			2050		
Areas	NCAR	GFDL	RCM	NCAR	GFDL	RCM
Farm practices in rain-fed areas	0.59	0.77	1.11	14.26	16.15	17.98
Additional irrigation	0.00	0.00	0.02	0.00	1.49	1.67
Total	0.59	0.77	1.13	14.26	17.65	19.65

Source: Authors' calculations based on data sources listed in table 3A.1.
Note: NCAR = National Center for Atmospheric Research; GFDL = Geophysical Fluid Dynamics Laboratory; RCM = Regional Climate Model.

Box 6.2 Unit Costs of Adaptation Interventions

Costing practices to adapt to climate change is challenging because of wide variation and the site specificity of unit costs. In general, particularly for capital-intensive interventions such as irrigation, the experience has been that costs in Nigeria are relatively high compared to similar countries in Africa and elsewhere. The analysis has therefore been conducted with regard to a range defined by a low unit cost case, and a high unit cost case. For irrigation, capital costs have varied from US$3,700/ha to $24,500/ha (JICA 1995). Recent consultations with irrigation experts in West Africa indicate an upper bound of US$5,000/ha for small-scale and $20,000/ha for large-scale schemes. Operation and maintenance (O&M unit costs used in this analysis) range from US$30/ha to US$40/ha.

For farming practices in rain-fed areas, the cost of applying manure and residues can vary based on the stocks available, as when livestock and cropping farms are integrated, as well as residues that can be reused in plowing. Among sources of information used are the Food and Agriculture Organization of the United Nations (FAO) study on low carbon development in Nigeria (FAO 2012) and two comprehensive reviews (Bationo 2004; Bationo, Waswa, and Okeyo 2011) that collect a series of papers on integrated soil management. We have mostly used the cost options given by the FAO, particularly for extension costs to cover changes in calendar years and the best selection of sowing dates and fertilization costs as integrated options to current practices. For the application of manure and residues, the source used is Mutiro and Murwira (2004) for the low-cost option; and Kamiri, Pypers, and Vanlauwe (2011) for the high-cost option.

Other irrigation options may cost more. Interventions to increase groundwater recharge, such as water harvesting or infiltration basins, may be needed as adaptation measures. Although not studied for this book because data were unavailable, groundwater irrigation and the necessary management measures should be part of any basin-wide water management plans to adapt to climate change.

In addition to direct outlays, there are also opportunity costs from diverting productive capital, which without climate change would have been allocated to other development priorities. The benefits are the value of the additional output that can be produced once adaptation measures are in place.

To evaluate the net effect, the macroeconomic model described in chapter 3 was run without negative climate change impacts on yields because these are fully offset by adaptation. The model run also incorporated a decrease in the annual capital stock because of the extra spending on adaptation. The metric used to assess the net effect is the terminal value of gross domestic product (GDP) in 2050, with adaptation and without. The results (table 6.5) indicate that, if unit costs can be kept in check, adaptation is effective at reducing net GDP loss.

In the low unit cost case, the terminal year GDP loss is always lower with adaptation than without; the benefit-cost ratio of adaptation ranges from 1.4 to over 3. But in the high unit cost case, the proposed adaptation strategy is no longer attractive: the opportunity cost of capital diverted to adaptation far

Table 6.5 Aggregate Adaptation Costs and Benefits

Variables	NCAR	GFDL	RCM
GDP loss induced by climate change in 2050	2.9%	3.6%	4.5%
GDP loss induced by adaptation in 2050			
Low unit cost case	0.93%	2.6%	2.3%
High unit cost case	5.15%	14.3%	12.7%
Benefit-cost ratio			
Low unit cost case	3.13	1.38	1.96
High unit cost case	0.56	0.25	0.35

Source: Authors' calculations based on data sources listed in table 3A.1.
Note: NCAR = National Center for Atmospheric Research; GFDL = Geophysical Fluid Dynamics Laboratory; RCM = Regional Climate Model.

exceeds the benefit in terms of recovered production. The benefit-cost ratio is consistently less than 1 in all climate scenarios.

The result underscores the importance of supporting adaptation with measures to control the unit costs of investments in irrigation and sustainable land management, which appear to be consistently much higher in Nigeria than in comparator African countries.

Robust Decision Making for Water Resources Infrastructure

The lack of agreement between different climate models on precipitation makes it difficult to project how much water will become available for storage. Reservoir size is typically designed to ensure sufficient storage to provide a given continuous or seasonal flow based on past weather patterns. However, historical data may no longer be adequate to guide long-term investment in water resources management and planning. If the climate changes, a storage volume identified on the basis of historical data may receive less water than expected and produce less benefit than projected, or there may be more water, and more benefit. Climate change impact must therefore be considered in the design of new projects of water storage and irrigation infrastructure development. The objective of this section is to evaluate what investment decisions on water resource management in general and on hydropower and irrigation development in particular are robust in a wide range of climatic outcomes—i.e., they are likely to deliver certain standards of irrigation or hydropower service in as many climate situations as possible. The risk of designing water infrastructure based on past weather patterns is quantified and a methodology for robust investment planning is presented. The section discusses surface water storage and irrigation infrastructure, but it is important to remind that all future investments in water need also to be analyzed as part of a basin or sub-basin system that takes into account other water users (including environmental flows) and increases in their demands resulting from climate change.

The hydrological modeling results described in chapter 3 have been used for this analysis, which is based on a comparison between the baseline (1976–2005),

Table 6.6 Climate Risks and Opportunities in Designing an Irrigation Dam

		Actual climate[a]	
		Dry	Wet
Expected climate at design stage	Dry (increased storage, reduced area)	No regrets	• Too much storage: regret from excess investment • Too little area: regret from foregone production
	Wet (reduced storage, increased area)	• Too little storage: regret from foregone production • Too much area: regret from excess investment	No regrets

a. Thus, it may be wise to prepare a worse and a better scenario for water availability in terms of both mean climate and year-to-year variability; i.e., drying future means here that Y2 < Y1 while S2 ≥ S1 or Y2 = Y1 while S2 > S1; wetting future means that Y2 > Y1 while S2 ≤ S1 or Y2 = Y1 while S2 < S1; where Y is the basin yield, S the storage, 1 and 2 indicate the first and second periods.

and 30-year future periods (2006–35 and 2036–65), simulated through the whole ensemble of climate projections.

The methodology proposed to make robust decisions consists in minimizing the regrets of adaptation decisions, i.e., of any decision to modify current practices in anticipation of future climate changes. Adapting a design to a future climate change has a certain cost, which is the extra capital cost of building storage or irrigated area. The cost becomes negative (turns into a benefit) if less storage or area needs to be built than the historical climate would direct. The benefit is the extra revenue obtained from selling hydropower or irrigated crops. The revenue becomes a potential loss if too little storage is built so that agricultural production is less or hydropower delivered lower. The regrets are defined as the difference in economic return between the chosen option ("no foresight") and the best possible option calculated for each scenario ("perfect foresight"). The regrets are illustrated in table 6.6 for the case of irrigation.

The study aim is to provide insights into irrigation and hydropower vulnerabilities as a way to inform policy decisions and orient development of technical capacity; although based on case studies, the results do not necessarily apply immediately to project design. Many assumptions had to be made about data and elements that are normally studied in detail at the planning stage. To inform actual project design, data used for the calculation should be updated and other elements incorporated, such as reservoir evaporation, risk of sedimentation, or the cumulative effects of multiple users. The range of climate models analyzed would probably also need to be broadened.

Irrigation

A robust decision-making approach to planning irrigation development was piloted in 18 planned dam sites to identify design options that could minimize regrets over a range of possible climate scenarios. Net present value (NPV) is the metric used to estimate the value of different investment decisions and calculate the regrets. Two objective functions were minimized, namely the average and the maximum regrets between scenarios, each reflecting a different degree of risk aversion. Optimizations were carried out with respect to either of two decision variables: the amount of stored water or the irrigated area. Then, if the purpose of the proposed dam is to irrigate a targeted area, the decision variable is the amount of storage. If the dam is already built or there are constraints on storage size, the decision variable is the size of the irrigated area.

Monthly inflow series from the sub-basin hydrological analysis make it possible to calculate storage-yield curves (SYCs) for each sub-basin. SYCs indicate the yield produced from a given level of storage, or alternatively the storage capacity needed to provide a given basin yield. In this study SYCs were built with reference to sequent peak algorithm (SPA; Thomas and Burden 1963), which is designed for studying reservoir capacity. The SYCs for future simulated flows show the combined effect of predicted changes in flow magnitude and inter-annual variability.

Figure 6.6 illustrates the steps used to optimize the amount of stored water needed to irrigate a targeted area. Eleven perfect-foresight storages (storage generating enough yield to provide water to the targeted irrigated area for a given climate scenario) are calculated for the RCM and 10 perturbation models. Then, one no-foresight storage is estimated and used to calculate a maximal irrigated area for each climate scenario. This may be larger or smaller than the target area. The difference in storage cost and irrigation revenues corresponds to the regrets and can be calculated for each scenario. The robust storage option is then obtained by finding the no foresight storage that minimizes the average and the maximum regrets for all scenarios. Similarly, robust decision on irrigated area can be estimated but the storage is assumed to be fixed while the irrigated area is optimized to minimize regrets.

This method is relatively easy to implement if climatic and hydrological data are available and can be used as a basis of discussion for decision makers. Additional details on methodology and assumptions are given in appendix J.

Figure 6.6 Analysis of Regrets in Irrigation

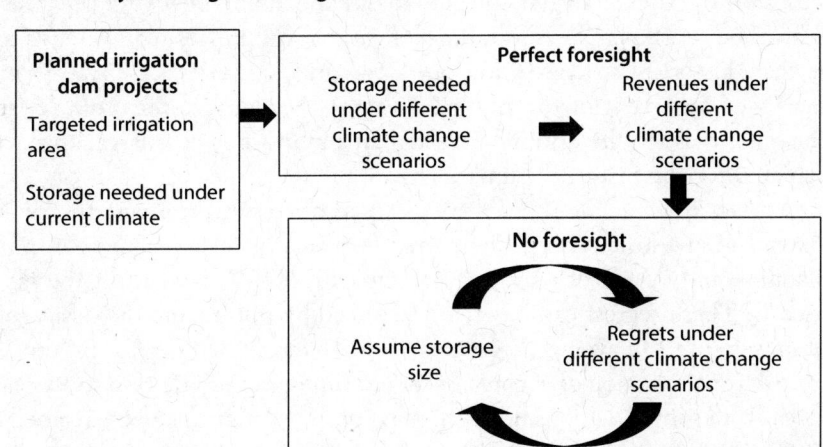

Map 6.1 Sites Selected for Robust Irrigation Decision Making

Site	Number
Ka	1
Zauro	2
Ajelanwa	3
Bakajeba	4
Kuda	5
Kunini	6
Hong Gombi	7
Ganye	8
Suntai	9
Karma	10
Leizi	11
Ambighir	12
Baushe	13
Oji	14
Ibu	15
Ogege	16
Moi	17
Yedesram	18

The case study sites (map 6.1) were selected in accordance with federal government plans as reflected in the Master Plan for Irrigation and Dam Development (2009–20) and using the following criteria: (a) the main basins where new irrigation development is planned should be represented; (b) the number of sites in each hydrological area (HA) should be proportional to the area planned for irrigation development in that HA; (c) catchment size should be larger than 100 km² (a sub-basin should represent the whole catchment); (d) there is no dam upstream; and (e) dry and wet future climates are represented. A small-scale irrigation dam in the dry HA8 was added (see appendix J for information on the schemes selected). This analysis attempts to illustrate the policy significance of the robust decision-making (RDM) approach. However, it should not be considered an assessment of the technical or financial feasibility of the design solutions investigated, which require more detailed investigation.

In most of the sites analyzed, the climate is expected to be wetter (i.e., the dry period will be shorter). As a result, the storage that minimizes regrets tends to be lower than historical storage. Only one case study requires significantly more storage (over 5 percent of the historical value) to minimize maximum regrets. For the same reasons, the optimal irrigated area in most cases is larger than what is calculated for the historical climate (figure 6.7).

The impact of changing the design was quantified by comparing the avoided regrets to the investment cost. The regrets of using historical climate as a basis for planning and design of irrigation are typically 10–40 percent of the investment cost.[2] These regrets can be greatly reduced by optimizing the design of an irrigation scheme. On average, regrets fall 30–50 percent depending on the aversion to risk (optimization of average or maximum regrets). The case study results vary greatly; in some locations up to 90 percent of the regrets can be avoided and the rest reduced by adding flexibility to the system: cropping patterns, water use,

Figure 6.7 Wet, Stable, and Dry Case Studies for Four Optimization Types

a. Minimizing maximum regrets: storage optimization

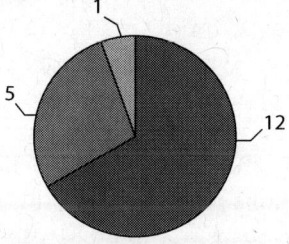

b. Minimizing maximum regrets: irrigated area optimization

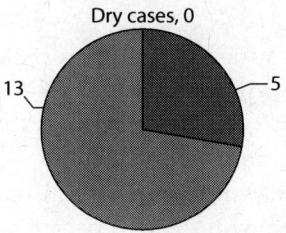

c. Minimizing average regrets: storage optimization

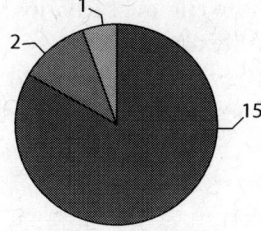

d. Minimizing average regrets: irrigated area optimization

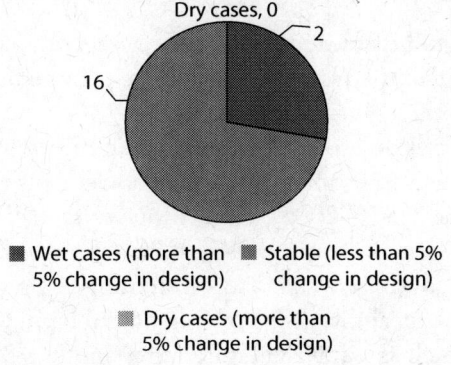

■ Wet cases (more than   ■ Stable (less than 5%
5% change in design)   change in design)

■ Dry cases (more than
5% change in design)

Source: Authors' calculations based on data sources listed in table 3A.1.

or other parameters can be adapted to wet or dry years to increase the return on irrigation investment.

Different classes of avoided regrets were defined based on their value compared to the investment cost. Optimizing the design has a high impact if the avoided regrets exceed 20 percent of the investment cost; the impact is low if avoided regrets go below 5 percent, and moderate if avoided regrets are between 5 and 20 percent. In about half the case studies, taking account of climate change in the design has a moderate to high impact whichever optimization method is considered (figure 6.8). Results obtained by optimizing the storage and the irrigated area are illustrated on maps 6.2 and 6.3, respectively.

The reduction in regrets exceeded 50 percent of the investment cost in two case studies in the northern part of the country (Ka in HA1 and Yedesram in HA8). There the climate is projected to be much wetter than the historical scenario for all the perturbed models, according to the mean annual runoff and the SYCs. Therefore, there is a strong incentive to build smaller dams to irrigate a given area, or define a larger irrigated area for a given storage. Nevertheless, these results should be viewed with caution because of the significant uncertainties in the climate and hydrological models.

There are several ways in which the method used here could be improved. In particular, to properly estimate SYCs, there is a need to keep in mind the accuracy of hydrological model simulations of dry conditions. Additional data on site characteristics should be collected locally to better estimate the economic return. It is equally important in calculating SYCs to consider the seasonality of irrigation demand, adding more climate scenarios in the analysis, integrating dam evaporation, or adding more flexibility to the system to adapt to wetter or drier conditions. Appendix J reports the full results of the site-level analysis, including a thorough analysis of the driest and the wettest sites.

Hydropower

In testing the robust decision-making approach for hydropower, because the economic analysis of different design options requires power simulation runs for every scenario, the methodology was simplified to compare only two types of design:

- Design according to historical records (base design)
- Design made to meet the worst (driest) future climate projection.

An alternative no-regret design for hydropower supply would aim to ensure that the frequency of failures to supply commercial power does not change compared to the baseline period (1976–2005), even if the worst of the 11 climate change scenarios simulated occurs. For the case study of the Mambilla hydropower scheme planned for southwest Nigeria, the amount of extra storage required to achieve the historical reliability is 5,310 million cubic meters for the worst climate projection, which is equivalent to adding another 31 m to the planned height, for a total of 1,330 m. However, since the maximum physically possible

Figure 6.8 Case Studies Where Adapting the Design Has a Low, Moderate, or High Impact on Regrets

a. Minimizing maximum regrets: storage optimization

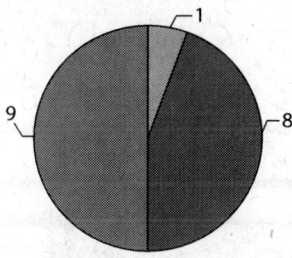

b. Minimizing average regrets: storage optimization

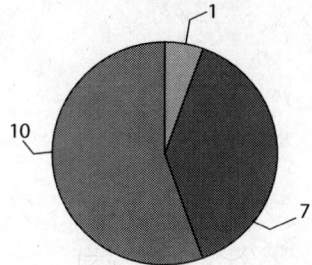

c. Minimizing maximum regrets: irrigated area optimization

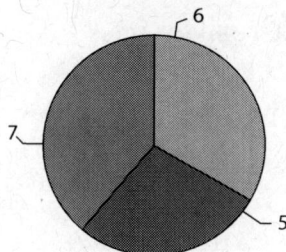

d. Minimizing average regrets: irrigated area optimization

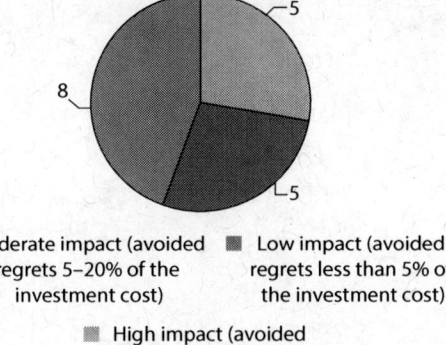

■ Moderate impact (avoided regrets 5–20% of the investment cost)   ■ Low impact (avoided regrets less than 5% of the investment cost)

■ High impact (avoided regrets more than 20% of the investment cost)

Source: Authors' calculations based on data sources listed in table 3A.1.

Map 6.2 Regrets Avoided by Optimizing Storage

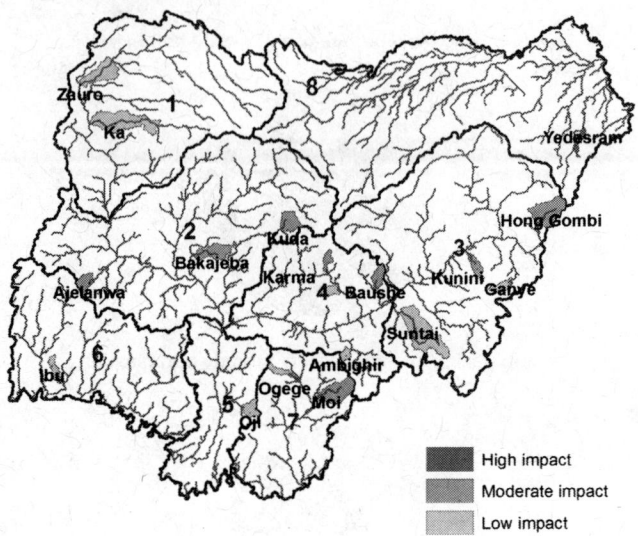

Source: Authors' calculations based on data sources listed in table 3A.1.
Note: Low impact: decrease in regrets is less than 5 percent of the investment cost; moderate impact: between 5 and 20 percent; high impact: more than 20 percent. The numbered units are hydrological areas.

Map 6.3   Regrets Avoided by Optimizing Irrigated Area

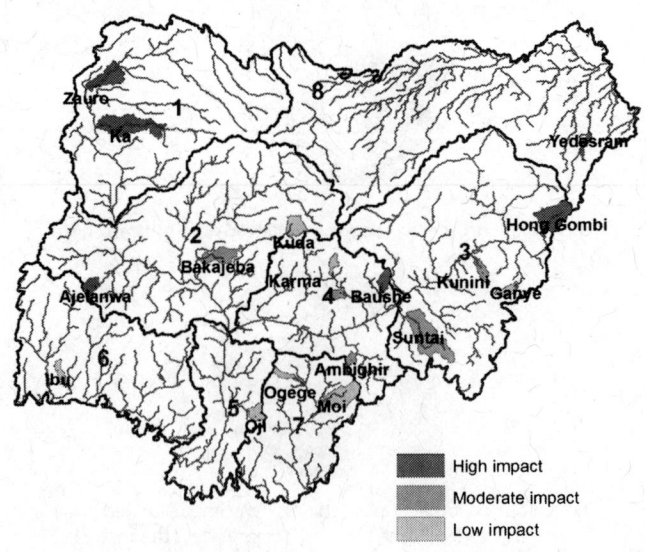

Source: Authors' calculations based on data sources listed in table 3A.1.
Note: Low impact: decrease in regrets is less than 5 percent of the investment cost; moderate impact: between 5 and 20 percent; high impact: more than 20 percent. The numbered units are hydrological areas.

dam height at Mambilla is 1,305 m, that figure was chosen as the most robust possible design. A dam height of 1,305 m gives an extra 1,200 million cubic meters storage above the base design level of 1,296 m (storage 2,900 million cubic meters). Design of a dam that is higher by 9 meters is considered low regret in terms of hydropower supply. It would cost US$110 million more than the optimum design according to historical data. This cost is compared to the cost of alternative energies calculated for all climate scenarios to identify the robust decision.

The basis for assessing the possible benefit of a robust design is that a higher dam may decrease the number of times the Mambilla will fail to deliver commercial power, since it will have more storage. Because replacing firm power is expensive, it may therefore be beneficial to accept a higher construction cost for Mambilla if that would reduce the cost of replacing power during the lifespan of the plant.

The analysis was based on the unit reference value (URV) of producing electricity for different climate change scenarios. URV is a measure of total unit cost per kilowatt hour (kWh) of firm power. When the hydropower plant is not able to produce the full firm power, it will be replaced by alternative power sources at higher costs. Alternative power sources could be (a) households using mid-sized diesel-generators to solve power deficits, and (b) a government guarantee that consumers do not suffer a power loss by using reserve power produced by a single-cycle gas turbine (SCGT). The URV is thus defined as the sum of

1. Capital costs for building the hydropower scheme
2. O&M costs for the hydropower scheme; and
3. The costs of any alternative power needed.

The more hydropower plants fail to deliver power, the more URV increases. It will also increase along with capital costs. Since the number of failures to deliver power will decrease if the storage dam is larger (meaning more capital costs), the URV works as an objective function that needs to be optimized. The ultimate URV is calculated from the consumer point of view (i.e., not from a profit-maximization point of view). The costs increase because a more expensive design will be transferred directly to consumers without subsidies. The URV was calculated for two scenarios for each dam design (table 6.7).

Table 6.7 Scenarios for the Evaluation of Unit Reference Value

Scenario	Source of back-up power	Levelized costs
Scenario 1: Inactive government: consumers have to solve power deficits themselves.	Cost of alternative electricity is calculated according to the price of diesel-generated power.	Levelized cost assuming mid-sized diesel generator (0.4 liter/kWh): 60 naira/kWh
Scenario 2: Active government: government guarantees that consumers will suffer no power deficit by bringing in reserve power from other sources.	Cost of alternative electricity is calculated according to the price of governmental gas-generated power plants.	Levelized cost assuming single-cycle gas turbine at 100 MW units: 13 naira/kWh

The URV calculation assumed 4 years for Mambilla construction followed by 50 years of operation. Because of the uncertainty in the hydrological model, two scenarios of conversions from climate projections to river runoff were further assumed. The results of the URV calculation for the whole ensemble of climate simulations and the regret for making the wrong decision are shown in table 6.8.

If diesel is the main alternative power source, the choice of the alternative design is economically beneficial in 3–5 of the 11 climate scenarios, depending on the hydrological model. Moreover, the regret of choosing the alternative design is lower than for choosing the base design. The maximum possible regret if the Mambilla scheme is designed according to historical data is 7–25 percent of the capital cost and maximum regret is only 5 percent with the alternative design, which is therefore the robust decision.

On the other hand, if the Government of Nigeria invests in gas-based schemes (SCGT) as back-up for replacing failure of the hydropower schemes to deliver firm power, the low-regret Mambilla design (higher dam) becomes less favorable. When the cheaper alternative power is available to replace the deficits, the alternative design is only beneficial in 2 of the 11 climate projections in the pessimistic hydrology scenario and in none using an optimistic hydrological model. The robust decision is therefore in this case to base the design on historical data and apply adaptive management by preparing alternative power from medium-expensive sources.

Analysis of robust design for future Nigerian hydropower schemes to take into account climate change therefore confirms careful planning as a powerful adaptation measure. A complete optimization to find the best solution, as conducted for irrigation dams, is possible to do as part of the preparation for a large infrastructure project like Mambilla. The example shows that designing only based on history without considering adaptation to climate change, may generate economic regrets. From the simple analysis of Mambilla, based on current climate projections, it seems that adaptation by providing alternative power through SCGT is preferred to a more robust dam infrastructure design. However, more detailed analysis would be needed to confirm this conclusion.

RDM in the Planning Process

The application of robust decision making for irrigation and hydropower sites in Nigeria has shown that it is possible to make climate change scenarios part of the planning process for water infrastructure. It is essential to note, however, that this methodology has to be applied case by case, and variables may differ depending on what is most sensitive for the economic viability of each case.

Because of data constraints, the applications of RDM for irrigation and hydropower sites in Nigeria were conducted using global hydrological data; they omitted such variables as dam evaporation and reservoir sedimentation, which may also be affected by future climate change. The intent was mainly to prove the benefits of using RDM as part of the planning process and to inform decision making. RDM should be applied in the detailed feasibility studies for water infrastructure to find the most optimal design. For feasibility studies, detailed

Table 6.8 Economic Analysis of Design Options for the Mambilla Hydropower Dam

Climate scenario	Pessimistic hydrology					Optimistic hydrology				
	A) URV for design based on historical records	B) URV for low-regret design	Best option	Regret if base design is chosen (discounted cost in U.S. millions)	Regret if low-regret design is chosen (discounted cost in U.S. millions)	A) URV for design based on historical records	B) URV for low-regret design	Best option	Regret if base design is chosen (discounted cost in U.S. millions)	Regret if low-regret design is chosen (discounted cost in U.S. millions)
10 percent discount rate and diesel-generated power replacing deficits										
CMCC-MED	0.662	0.702	A		$89	0.662	0.702	A		$89
CNRM	0.662	0.702	A		$89	0.662	0.702	A		$89
CSIRO	0.869	0.702	B	$369		0.833	0.780	B	$118	
GFDL	0.741	0.712	B	$65		0.741	0.754	A		$30
IAP	0.662	0.702	A		$89	0.662	0.702	A		$89
MIROC	0.662	0.702	A		$89	0.662	0.702	A		$89
MPI	0.749	0.702	B	$103		0.749	0.704	B	$100	
MRI	0.945	0.746	B	$440		0.945	0.886	B	$130	
NCAR	0.662	0.702	A		$89	0.662	0.702	A		$89
RCM	0.671	0.702	A		$62	0.671	0.703	A		$72
UKMO	0.728	0.702	B	$56		0.727	0.730	A		$6
Number of scenarios where option is better	6	5				8	3			
Maximum regrets				$440 (25% of capital cost)	$89 (5% of capital cost)				$130 (7% of capital cost)	$89 (5% of capital cost)

table continues next page

Table 6.8 **Economic Analysis of Design Options for the Mambilla Hydropower Dam** *(continued)*

	Pessimistic hydrology					Optimistic hydrology				
Climate scenario	A) URV for design based on historical records	B) URV for low-regret design	Best option	Regret if base design is chosen (discounted cost in U.S. millions)	Regret if low-regret design is chosen (discounted cost in U.S. millions)	A) URV for design based on historical records	B) URV for low-regret design	Best option	Regret if base design is chosen (discounted cost in U.S. millions)	Regret if low-regret design is chosen (discounted cost in U.S. millions)
10 percent discount rate and single-cycle gas turbine replacing deficits										
CMCC-MED	0.662	0.702	A		$89	0.662	0.702	A		$89
CNRM	0.662	0.702	A		$89	0.662	0.702	A		$89
CSIRO	0.707	0.702	B	$10		0.699	0.719	A		$44
GFDL	0.679	0.704	A		$56	0.679	0.713	A		$76
IAP	0.662	0.702	A		$89	0.662	0.702	A		$89
MIROC	0.662	0.702	A		$89	0.662	0.702	A		$89
MPI	0.681	0.702	A		$48	0.681	0.702	A		$48
MRI	0.723	0.712	B	$26		0.723	0.742	A		$42
NCAR	0.662	0.702	A		$89	0.662	0.702	A		$89
RCM	0.664	0.702	A		$85	0.664	0.702	A		$85
UKMO	0.676	0.702	A		$58	0.676	0.708	A		$71
Number of scenarios where option is better	9	2				11	0			
Maximum regrets				$26 (1.4% of capital cost)	$89 (5% of capital cost)					$89 (5% of capital cost)

Source: Authors' calculations based on data sources listed in table 3A.1.

Note: URV is unit reference value expressed in U.S. cent/kWh power. The URV only covers part of the production cost related to the capital cost and operation and maintenance of the scheme, and does not represent the true cost per kWh. Capital cost for base case design of dam is US$1,800 million and for robust design US$1,910 million. URV is based on 50 years of production plus cost of alternative power production from diesel or SCGT when the Mambilla dam fails to deliver firm power. CMCC-MED = Euro-Mediterranean Center on Climate Change; CNRM = Centre National de Recherches Météorologiques; CSIRO = Commonwealth Scientific and Industrial Research Organization; GFDL = Geophysical Fluid Dynamics Laboratory; IAP = Institute of Atmospheric Physics; MIROC = Center for Climate System Research; MPI = Max Planck Institute; MRI = Meteorological Research Institute; NCAR = National Center for Atmospheric Research; RCM = Regional Climate Model; UKMO = United Kingdom Meteorological Office.

data would be available and the RDM should take into account decision variables that may be used to attenuate the effects of climate change cost-effectively.

Notes

1. So, for example, if for a given crop, AEZ, and climate model, adaptation option A delivers a 10 percent yield increase compared to the no-adaptation case; option B a 15 percent improvement; and the best option (for the same combination of crop, AEZ, and climate model) delivers a 20 percent improvement; then the "regret" of adopting option A rather than the best option of its group is A's performance gap compared to the optimum, i.e., 50 percent; in the case of option B the regret is 25 percent.

2. The regrets can be as high as 114 percent of the investment cost for the Yedesram case study, but as explained later, this result should be viewed with some skepticism.

References

Bationo, A., ed. 2004. *Nutrient Cycle to Sustain Soil Fertility in Sub-Saharan Africa*. Nairobi: Academic Science Publishers.

Bationo, A., B. S. Waswa, and J. M. Okeyo, eds. 2011. *Innovations as Key to the Green Revolution in Africa*. New York: Springer.

FAO (Food and Agriculture Organization). 2012. "Low-Carbon Development in Nigeria's Agriculture Sector." Background report, FAO, Rome.

JICA (Japan International Cooperation Agency). 1995. "The Study on the National Water Master Plan" (Sector Report Vol. 2). Report prepared for the Federal Ministry of Water Resources and Rural Development, Abuja, Nigeria.

Kamiri, W. M. H., P. Pypers, and B. Vanlauwe. 2011. "Residual Effects of Applied Phosphorus Fertilizers on Maize Grain Yield and Phosphorus Recovery from a Long-Term Trial in Western Kenya." In *Innovations as Key to the Green Revolution in Africa*, Vol. 1, edited by A. Bationo, B. Waswa, J. M. Okeyo, F. Maina, and J. Kihara. New York: Springer, 717–27.

Mutiro, K., and H. K. Murwira. 2004. "The Profitability of Manure Use on Maize in the Small-Holder Sector of Zimbabwe." In *Managing Nutrient Cycles to Sustain Soil Fertility in Sub-Saharan Africa*, edited by A. Bationo. Nairobi, Kenya: Academic Science Publishers, 571–82.

Thomas, H. A., and R. P. Burden. 1963. *Operations Research in Water Quality Management*. Division of Engineering and Applied Physics, Harvard University, Cambridge, MA.

CHAPTER 7

Conclusions and Recommendations

The findings of the analysis, summarized in box 7.1, indicate that in Nigeria, as in many other countries, climate change impacts will be significant, especially in the medium to long term. There are at least three reasons why the government may wish to act now to address them:

1. Many actions that will build up longer-term climate resilience will also help reduce vulnerability to current climate swings.
2. Decisions that will be taken in the near future will determine how resilient investments in long-lived (and expensive) infrastructure, such as irrigation or hydropower, will be to the harsher climate of the future.
3. Building the knowledge, capacity, institutions, and policies needed to deal with the climate of the future takes time. The longer that Nigeria delays action, the less time it will have to get ready, and the more it will have to resort to cure, which is typically more expensive and less effective, rather than prevention.

There are a number of actions that Nigeria could consider to enhance its ability to plan and put in place climate-resilient development. These can be organized around the three areas of institutions, information, and investments. Table 7.1 provides a synoptic view of the recommendations discussed in the rest of this chapter and an indication of when action is proposed: over a shorter time span (say 2015); and over a longer one (after 2015).

Institutions and Policies

Governance Framework

Nigeria is a party to the UN Climate Change Convention (UNFCCC); has ratified the Kyoto Protocol; and adheres to the Copenhagen and Cancun Accords and the Durban Platform. In 2003 Nigeria submitted its first national communication to the UNFCCC but has not yet completed the second one. On the domestic front, the Federal Ministry of the Environment has taken steps to move the climate agenda forward; it has established an interministerial committee for

Box 7.1 Main Findings of the Impact and Adaptation Analysis

- *Climate projections:* The significance of climate shifts will increase in the medium term (2036–65) compared to the short term (2006–35).

   Results from climate analysis indicate that on average temperatures in Nigeria will rise from 1°C to 2°C, with the already arid north more affected than the wetter south. Changes in the amount of projected rainfall were not particularly evident for the nation as a whole, and there was no clear agreement whether rainfall would rise or fall.

   The spatial variability of rainfall projections is largely higher than the projections for temperature, and shows sub-regional patterns that may be particularly important for state planning activities.

   The combination of changes in temperature and precipitation show biophysical impacts that can have significant consequences for the agriculture and water sectors and the hydropower subsector. The likely negative impacts of climate change on rain-fed agriculture and livestock and the increased uncertainty about water resources available in the future make it essential to factor climate change into development in the water, agricultural, and energy sectors.

- *Crop yields:* Agriculture, both crop and livestock, will mainly be affected by increasing loss of yields for the main crops (cassava, millet, yam, maize, sorghum, and rice), even if precipitation increases in several parts of the country. The effects are more uncertain in the shorter term (2020), when, according to more than half of the climate models, cassava and perhaps other crops might actually experience an increase in yields.

- *Food security:* The projected decline in rain-fed yields along with projected rises in temperature might ultimately reduce food security. It is projected that half of Nigeria's agroecological zones (AEZs) will be food insecure by 2020 and 75 percent by 2050 unless their dwindling local food production is complemented by improved in-country trade (which requires better infrastructure for transport and storage) or more imports.

- *Livestock:* The temperature increase projected is likely to cause stress on livestock productivity and trigger higher mortality rates. Along with reduced yields from rain-fed crops, this trend is likely to have serious implications for livelihoods and poverty in Nigeria.

- *Water resources:* Impacts on water resources are more uncertain, but it appears that availability of water for storage and use will be different than in the past. Disagreement among climate models on precipitation makes it difficult to project how much of rainfall could eventually be used for water supply, irrigation, and hydropower schemes.

   The analysis suggests that by 2050, in about half of the country wet risk is expected be dominant, 10 percent of the country to be exposed to drier conditions, and 23 percent to be stable—but 33 percent of total land area is subject to uncertainty. The projected change in water availability shows that caution must be used in allowing historical records to guide decisions about future investments.

- *Macroeconomic impacts:* The decline in crop yields will have significant consequences for the national economy, by 2050 reducing GDP (compared to the no-climate change scenario) by up to 4.5 percent. Climate change is also projected to increase net import of various crops, particularly rice and other cereals.

box continues next page

Box 7.1 Main Findings of the Impact and Adaptation Analysis *(continued)*

- *Adaptation options in agriculture:* There is a wide range of land and water management practices that can offset or even reverse the effects of climate change on crops and livestock. Many are aspects of conservation agriculture (e.g., integrated soil fertility management, water harvesting, agroforestry). Other options are shifts in sowing/planting dates, crop rotation, minimum or no tillage, and restoration of degraded pasture.

  A combination of robust sustainable land management practices for 14–18 million hectares (ha) of rain-fed areas and 1.5–1.7 million additional irrigated ha might fully offset long-term climate change impacts on agriculture. At low unit costs, this adaptation package has a benefit-cost ratio exceeding 1 in all climate scenarios considered.

- *Adaptation options for water resource management:* Irrigation is important both for baseline development and as an adaptation strategy. But if climate history is used in designing new projects, the investments may be under- or over-sized. By applying the robust decision-making approach to 18 real-life projects, this book finds that the regrets for not including climate change in the design can be as high as 40 percent of investment costs; and that by selecting the investment strategy that minimizes regrets across multiple climate outcomes, they can be reduced by 30–50 percent on average, and up to 90 percent in some locations.

  Climate change does not affect the feasibility of exploiting Nigeria's hydropower potential. On grounds of the energy diversification and low carbon co-benefits, taking advantage of the full 12 gigawatts (GW) of potential should be considered. However, uncertainty about future precipitation and river runoff makes it challenging to optimize the economic effects of hydropower schemes.

  Here, too, robust decision making could be considered. For example, a first-cut analysis of the planned Mambilla scheme indicates that because of the possibility of a drier climate there, the project risks not delivering the intended amount of power. Under certain assumptions, designing the dam without taking into account climate change exposes the project to a regret (the cost of failing to deliver power) of 25 percent of capital costs; using a robust approach to design that increases storage in anticipation of a possibly drier climate reduces the regrets to 5 percent of capital costs.

---

climate change and set up a special climate change unit, recently upgraded to a department.

To consolidate progress, the National Assembly recently passed a bill to establish a National Climate Change Commission (NCCC), which is mandated to coordinate national policies on climate change. At the time this book was prepared, the bill was still awaiting final endorsement by the president.

Pending a decision on activation of the NCCC, there is a need to move the climate resilience agenda forward along two paths:

1. Elevate the policy profile of the agenda. The Federal Economic Management Team could be charged with defining priority adaptation actions, building on the 2011 National Adaptation Strategy and Plan of Action on Climate Change

Table 7.1 Recommendations by Area and Time Horizon

Action through 2015, results by 2020		Action after 2015	
Recommendation	*Lead agencies*	*Recommendation*	*Lead agencies*
A. Institutions and policies			
Define priority adaptation actions for each sector	Economic management team Federal Ministry for the Environment		
Harmonize policies and legislation related to water resources management	Federal Ministry of Water		
B. Information and knowledge			
Launch a dedicated program of applied research and extension on climate-smart agriculture (CSA)	Federal Ministry of Agriculture	Improve access to and sharing of publicly funded data among federal and state ministries, departments, and agencies	Federal and state ministries of agriculture, NASRDA, NIHSA, NIMET
Define an action plan for building up extension services	Federal Ministry of Agriculture, states, commercial service providers, producer associations, religious organizations, nongovernmental organizations	Develop a south-south cooperative program with countries like Brazil, India, and China	Federal government
Create planning tools for CSA (e.g., a CSA atlas)	Federal Ministry of Agriculture	Recognize and reward farmers through small grants, competitions, and media exposure	States
Draw up an action plan for better monitoring the hydrometeorological system	NIHSA (Federal Ministry of Water)	Strengthen the hydrometeorological system	Federal government
Draft guidelines for designing climate-smart water infrastructure	Federal Ministry of Water	Analyze climate risks in Lagos and identify and prioritize adaptation measures	Lagos state government
C. Investment and resource mobilization			
Include in the Agricultural Transformation Agenda (ATA) a dedicated program of CSA demonstration projects	Federal Ministry of Agriculture	Improve incentives for adoption of CSA measures	States
Incorporate robust decision making in the feasibility studies for individual irrigation and hydropower projects	Federal Ministry of Water; Federal Ministry of Power	Integrate the no regret/ robust decision-making analysis into current investment operations, particularly in irrigation	States and federal government
Set up in a few states integrated watershed management and monitoring plans	Federal Ministry of Water and para-statals, Federal Ministry of Agriculture, Federal Ministry of Environment, state governments	Budget more spending for extension services	State government

for Nigeria (NASPA-CCN), and on the results of the present analysis. This would ensure that enhancing climate resilience becomes a cross-cutting priority, not just a concern of the Ministry of the Environment; and that there are clear directions for coordinating, across institutional boundaries, the climate-related actions of different ministries, departments, and agencies (MDAs). The priority adaptation actions should include significant efforts—to be sustained over time—to increase capacity, to ensure that climate resilience becomes part of the core competencies of relevant staff in MDAs.

2. Ensure that national policy decisions are adequately informed by technical work. While much of that work needs to take place within federal and state line ministries (as well as locally in communities), as suggested by the experience of China (see box 7.2), there is a need for an institutional "champion" that can promote integration of the climate agenda, across sectors and ministries, departments, and agencies at all three tiers of government. Such a champion needs to have adequate technical credibility to provide timely and practical expertise and advice to partner MDAs in climate-related decision making.

The champion role can be played by the Federal Ministry of the Environment, although it needs to be reinforced in terms of capacity and resources; or by the proposed NCCC, should it be approved and be assigned a technical role on top of the policy one. Functions that should be discharged at the central level include:

- Collect, analyze, and act upon data and information on climate risks and low carbon development. Examples of priority information are remote-sensed data on soil moisture and vegetation cover, hydrological data, weather data, crop data, and household survey data. Inter-MDA memoranda of understanding and joint work programming overseen by the country's lead climate change authority, once such an authority is defined, could kick-start this process. The latest project in the Fadama series includes a budget for promoting data collection and knowledge sharing and can be deployed immediately.
- Facilitate integration of climate considerations into policies and development planning.
- Mobilize national and international resources for climate action.

In sum, a climate change champion would need to provide real-time services to its partner MDAs to help make the Nigerian economy more climate-resilient and less carbon-intensive than it currently is.

Water Sector Institutions

In addition to the need to address cross-cutting and cross-sectoral aspects, making progress on the adaptation agenda requires reinforcing institutions in specific sectors. A particularly important area is water resource management, since those investment decisions will be particularly affected by climate change.

Box 7.2 Adaptation of China's Agriculture to Climate Change

For the past 50 years, the Huang-Huai-Hai (3H) Plain, a major agricultural area in China that is critical to the country's economy and its food security, has seen a clear climate warming trend. The mean temperature has risen by 1.18°C, and annual mean rainfall fell by 140 mm between 1954 and 2000, causing more frequent spring droughts, with severe effects on crops. By 2030 much of the region could face a serious water deficit.

In 2004, the Chinese government responded by launching a World Bank–financed project that worked with farmers and technical experts to implement water-saving measures in the five provinces in the plain. The Ministry of Finance's State Office of Comprehensive Agricultural Development (CAD) coordinated activities with assistance from the ministries responsible for water resources, agriculture, land, and forestry.

The goals were to reverse the inefficient use of water for farming and to increase the financial returns to farmers. From 2005 to 2010, irrigation-centered engineering, agronomic, and management measures were implemented at a cost of US$463 million across 107 counties. The target was to improve 505,505 ha of low- and medium-yielding farmland, benefiting 1.3 million farm families. The formation of water users' associations, encouraged by the government, provided forums for training in new water management techniques, as well as a mechanism for better local water management based on farmers' participation. Irrigation facilities constructed as part of the project were handed over to these associations so that farmers could manage and maintain the infrastructure.

Since the project's design did not systematically integrate the risks posed by climate change into all its activities, in 2006 CAD requested, and received, a grant from the Global Environment Facility (GEF) to incorporate adaptation activities into the World Bank–supported irrigation and agricultural program.

It took several growing seasons for many farmers to adopt the new crop varieties, but their reluctance to shed their decades-long reliance on what one villager called "the same old wheat" was eventually overcome by the higher yields delivered by the new varieties. Similarly, government-led pilot programs introducing new techniques to better manage irrigation water took hold after the farmers saw the benefits for themselves in less waste water, cheaper irrigation, and reduced groundwater depletion—all resulting in greater water productivity.

A critical condition for adopting and up-scaling this approach proved to be the introduction of a credible and well-equipped coordinating agency, to ensure the smooth implementation of adaptation measures, including a well-established institutional design, a good reputation among farmers, and a continuing investment program that combines development of both infrastructure and software.

Source: World Resources Institute 2011.

The institutional configuration of the sector is still evolving. In 2007 the government established the Nigerian Hydrological Services Agency (NIHSA) as a response to the need to provide reliable hydrological data for developing water resources. Also in 2007, the Federal Executive Council approved establishment of the Nigeria Integrated Water Resources Management Commission (NIWRMC).

The Commission is tasked to regulate and control the rights of all actors to develop and use water resources shared by more than one state. The current governance structures do not always facilitate adaptation to prevent or mitigate the effects of climate change in Nigeria. One example is the legal and institutional framework related to water resources and disaster management, where different authorities have overlapping and possibly inconsistent responsibilities.

The recent 2007 Water Resources Bill contains promising measures based on the concept of integrated water resources management, such as the establishment of a National Council on Water Resources and eight River Basin Management Commissions. The bill is currently being reviewed to strengthen some provisions, such as the relationship with state and local governments, financing, and monitoring and evaluation.

It is essential that the Federal Government accelerate the reform process already begun, to consolidate and harmonize policies, legislation, and institutions related to water resources management, as a prerequisite to organic and effective integration of climate change considerations into sector planning and development.

One vehicle for support for that agenda is the new World Bank–financed Nigeria Erosion and Watershed Management Project (NEWMAP), a multisector operation which brings together the MDAs already mentioned to prioritize actions using an integrated watershed approach that balances the competing needs of land and water users. This work complements activities being carried out throughout Nigeria by various MDAs.

Information and Knowledge

National Agricultural Research System

Nigeria's complex National Agricultural Research System (NARS) encompasses a large number of stakeholders. Recent research (box 7.3) has identified strategies to enhance the system's capacities.

To make progress on the climate change agenda, the Federal Ministry of Agriculture could launch a dedicated applied research program on climate-smart agriculture (CSA), with individual research lines to be awarded competitively to institutions in the NARS. The program could:

- Formulate dedicated programs on climate adaptation. Knowledge efforts could focus on producing planning tools (e.g., a CSA atlas to map land suitability for agricultural development as the climate changes) to define and prioritize, across space and value chains, opportunities for adopting "triple-win" agricultural options (higher yields, higher climate resilience, lower carbon emissions); analysis of implications of climate variability and change on specific value chains, including some of those with high-value crops, such as cocoa, that have considerable export potential but could not be addressed in this book. The efforts could also give priority to defining solutions in the field that farmers can adopt.

Box 7.3 Improving R&D and Innovation in the Agricultural Sector

Nigeria has arguably the largest and most complex National Agricultural Research System in Sub-Saharan Africa, incorporating a wide network of university agronomic and veterinary sciences departments and the facilities of the CGIAR (Consultative Group on International Agricultural Research). A recent IFPRI report (2010) assesses the innovation capacity of the Nigerian agricultural research system and discusses options for strengthening it:

- *Improve collaboration between researchers and promote communication on innovations.* Although research productivity seems high, the level of collaboration is low and there is no clearly structured monitoring and evaluation of the use, influence, and impact of technologies and publications being produced by organizations and individual researchers.
- *Increase interactions with farmers, the private sector, extension agents, and other actors within the system.* Greater awareness and sensitization, as well as exposure to practical knowledge, good practices, and experiences on innovation systems in other countries, are urgently needed. The Agricultural Research Council of Nigeria can facilitate a platform or forum for greater interaction and collaboration.
- *Strengthen fundraising abilities and diversify fund sources.* The agricultural research organizations have substantive capacity and incentive gaps. Among research institutes, the timely release of funds is the top motivating factor researchers identified as necessary to produce more and be more innovative.
- *Improve governance of research organization*: Good performance and innovation capacity are associated with fair and transparent hiring procedures; effective performance evaluation and reward systems; systems of career development and job security; systems of information sharing and knowledge management; clearly defined and communicated division of roles and responsibilities; systems of feedback from stakeholders; and flexibility, freedom to do work, and mobility for researchers.
- *Establish a mechanism for continuous training and skill development.*

Concerning the implementation of such measures, the recent World Bank review (2008) of public spending on agriculture suggests that *improving the quality of public spending in agriculture* could deliver greater benefits than could be achieved by simply increasing the amount of public spending without revising the composition of outlays and increasing efficiency. Rigorous external evaluation is needed to assess public program performance, for example for the National Special Program for Food Security, and generate information on which to base design adjustments. Likewise, very little is known about the impact of public support on fertilizer in Nigeria: What crops have benefited from fertilizer programs, and who grows those crops? What has been the impact of increased fertilizer use in terms of productivity increases, income growth, and poverty reduction?

Source: IFPRI 2010.

Priority climate-smart practices for analysis would be improved seed varieties, change in seeding dates, minimum tillage, effect of climate variability and change on the quality of high-value crops like cocoa, natural regeneration and agroforestry, grazing land management regimes, such as rotational grazing, integrated soil fertility and nutrient management, and rainwater harvesting.

- Enter into a south-south cooperative program with countries like Brazil, India, and China, involving relevant private sector agents. Bank-financed projects such as Fadama and NEWMAP all have budgets for international knowledge exchange activities.

Extension Services

The Federal Ministry of Agriculture could define an action plan for building up extension services through partnerships and cost-sharing arrangements in 5–10 states, such as helping farmer organizations to access carbon markets. The plan should be backed up by an agreement with the Ministry of Finance to provide enough federal resources to ensure sustained functioning of extension over time. In particular, this action plan could

- Promote collaboration between researchers, extension agents, and farmers in order to communicate and disseminate innovation; to heighten researcher incentives, sources of funding should be diversified and transparent.

- Recognize and reward farmers through small grants, competitions, and media exposure for farmers and communities that are putting in place successful and innovative adaptive measures. It is important to leverage the capacities of these farmers and communities by empowering them to disseminate their knowledge to their neighbors, e.g., through support by state extension agents, NGOs, and religious groups. Likewise, recognize and reward trained extension agents for their efforts to mobilize, equip, and empower their client communities.

- Build up the capacity of the country's extension services (Agricultural Development Programs; ADPs) to provide real-time expert advisory services on climate resilience and low-carbon technologies. This should include strengthening on-the-ground coordination between agricultural and forest extension personnel, given the challenges at the forest/agriculture frontier that have the potential to reduce sector performance of agriculture, forest, and water if not addressed holistically.

- Provide a platform to scale up participatory farmer-to-farmer learning and farmer champions. It is often difficult to identify well-connected and credible champions who will host on-farm demonstrations and learning events that are critical for scaling up, but this is typically central to any strategy to scale up specific technologies.

- Target commercial service providers for strategic value chains. In areas that government cannot yet reach, the private sector is already present. It is important to demonstrate the value climate-resilient technologies can add to the quality and quantity of production, especially for high value-added crops like cocoa.

- Given the seriousness of the impact of climate change on livestock vulnerability, give priority to building up extension services to pastoralists, both nomadic and sedentary. Practices to be promoted could include rotational grazing, scaling up the use of grazing corridors, use of vegetation cover to ameliorate heat stress, establishing watering points, and mechanisms to reduce conflicts about natural resources management (using the conflict-reduction measures proven in the Bank-financed Fadama operation as one possible model).

Water Data

After 1990 Nigeria did not produce any hydrological data because of lack of funding. Despite the effort in 2007 to restore the hydrological monitoring network and its operation, due to limited financial resources and unclear allocation of responsibilities between federal, regional, and state authorities, in 2011 Nigeria still fell short in producing reliable river runoff data from more than a handful of stations. According to the recommendations of the World Meteorological Organization (WMO), a country the size of Nigeria should have at least 500 functioning runoff stations to enable good planning for water resource development. The lack of more than 20 years of hydrological data is a serious problem for putting in place sustainable water resources measures, such as dams and other hydraulic infrastructure. It also prevents efficient flood and disaster management.

The Nigerian Meteorological Agency (NIMET) currently operates 40–50 meteorological stations in Nigeria, although WMO recommendations for rainfall monitoring set the minimum number of gauges networked in Nigeria at more than 1,500. Based on the findings of this analysis, it is important to

- Build up the hydrometeorological system, in terms of both the density of the observation network and enhanced headquarters and river basin agency capacity to organize and analyze data for decision making. A reasonable target for increasing current station density by 2020 might be 30–50 percent. It is also necessary to
  - Ensure that data are available to a variety of ministries and agencies: understanding climate change requires collecting and analyzing data from a wide range of sources (in many states, these include agencies operating in the energy, agricultural, or water sectors); the latest data should be provided free or at affordable prices in the medium and long terms through agreements between ministries and agencies.
  - Learn more about groundwater use and users: groundwater is a key resource for buffering climate change; smallholders already use it for irrigation. But actual borehole distribution and yield from boreholes cannot be surveyed, which makes it difficult to take action against over-pumping and groundwater contamination. Current law cannot regulate lowering of groundwater levels, or seawater intrusion and groundwater contamination by illegal injection of industrial waste water into boreholes (JICA 2012).

- Draw up guidelines to optimize the design of climate-smart infrastructure for water storage and use, taking into account (as discussed in chapter 6) the full range of possible climate outcomes, so that hydropower and irrigation schemes are able to meet standards of service, no matter what the future climate will be.

Investment and Resource Mobilization

The Agricultural Transformation Agenda

The Federal Government's Agricultural Transformation Agenda (ATA) is a key component of its reform program. The agenda purports to bring about a major overhaul of the sector (box 7.4), focusing on enhancing the productivity of several supply chains, including various crops included in the analysis of this book.

The ATA could be the natural vehicle for promoting early experimentation with triple-win solutions—higher production, higher climate resilience, and enhanced carbon storage—as a way to create a platform to then scale up practices that over time will have effects in the field. The experience in other countries in West Africa (box 7.5) indicates that such efforts eventually pay off; but success needs time, so there is a premium attached to early action.

The government could add to the ATA a dedicated program of demonstration projects on CSA to pilot and scale up climate-resilient technologies in different agro-climatic settings. Based on the findings of the adaptation analysis reported in chapter 6, the government could consider establishing a target of up to 1 million ha under sustainable, climate-resilient land management practices by 2020. This is the lower bound of the area required to offset shorter-term climate

---

Box 7.4 Nigeria's Agricultural Transformation Agenda

The 2012 Agricultural Transformation Agenda (ATA) is a comprehensive plan to restore Nigeria's old glory as an agriculture powerhouse. It seeks dramatic increases in agricultural productivity, massive agricultural job creation, significant expansion of value-addition in processing, drastic reductions in agricultural imports, and deeper penetration into international markets. Among the commodities it targets are rice, cassava, cacao, oil palm, cotton, sorghum, maize, soybean, tomato, onion, livestock, and aquaculture, differentiated by Nigeria's six geopolitical zones.

The ATA is the point of departure for transforming Nigeria's agriculture sector by providing (a) in-depth analysis of root causes of poor performance of the agriculture sector and quantification of lost opportunities caused by this poor performance; (b) a clear vision for transformation of the sector as a process, including import substitution, export orientation, and value-addition through processing and backward integration linkages; (c) an explicit focus on agriculture as a business, putting the private sector in the driver's seat and recognizing the critical role of women; (d) a comprehensive approach to change by focusing on value chains; (e) a concrete and specific program of sector policy reforms, including revamping of the fertilizer subsidy program, which has been a major financial drain; and (f) quantified targets for expected outcomes in terms of jobs, income, food security, and productivity improvements.

Box 7.5   Assisted Natural Land Regeneration in Burkina Faso and Niger

The degradation of land is a major problem in Sub-Saharan Africa. Numerous direct and indirect drivers, such as population increase, growing demand for resources, poverty, and lack of effective governance, all potentially exacerbated by climate change, are reducing the productivity of land and hence threatening food security and the lives of millions of people.

Sustainable forest management (SFM) in dry lands is directed to reversing land degradation and maintaining food security in the long term. Forests are also significant factors in regulating the global climate. In Burkina Faso and Niger, one SFM approach, assisted natural regeneration of degraded land, has successfully addressed issues, such as soil erosion, fertility decline, sealing and crusting, reduction of vegetation cover, and aridification.

As example, 3 ha of degraded land were enclosed with a solid fence and a dense living hedge of thorny trees. A strip of 10 m along the hedge was dedicated to agriculture, the rest to natural regeneration of the local forest. Once protected, natural vegetation rich in endogenous species actively regenerated and biodiversity evolved. The forest reached a density of about 500 trees per hectare. Management in the protected area included (1) seeding/planting of improved fodder species; (2) establishing stone lines and half-moons to control erosion and harvest water; (3) installing beehives for honey production; and (4) producing fodder.

The success of assisted natural regeneration can be attributed to several factors. For instance, property rights for the protected area were established through a contractual agreement that incorporated traditional and government land rights. Training was provided to enhance income-generating activities, from beekeeping and production of high-value vegetable crops to processing of nontimber forestry products, and to promote use of fuel-efficient cooking stoves.

The long-term benefits of the approach are numerous, among them economic benefits from higher crop yields, ecological benefits from better soil cover, and sociocultural benefits from improved food security. Assisted natural regeneration of degraded land has huge potential for scaling up. The greatest benefits arise from strategic combinations on the same plot and scale up to landscape level.

Critical conditions for adopting and upscaling this approach are to

- Raise awareness of SFM
- Support pioneering communities
- Provide training
- Introduce inventories and long-term monitoring
- Ensure a favorable enabling environment through helpful laws and institutions
- Capitalize on opportunities to add value.

Source: Liniger *et al.* 2011.

impacts on agriculture. The proposed program could focus on regions particularly vulnerable (e.g., in the north and in the south-west); and on strategic crops and supply chains, such as rice, which appears vulnerable in many climate scenarios, and cassava, which at least in the medium term may be better suited to cope with a changing climate.

Other actions that the federal government could consider might be to

- Improve the enabling environment for responsible private investment, by exploring innovative agricultural financing through public-private partnerships. Examples are bundling agricultural credit and insurance together, and providing different forms of risk management, such as index-based weather insurance or weather derivatives.
- Set up a program of technical assistance to farmer organizations to enable trading, in the voluntary and compliance markets, of carbon assets developed as a result of implementing CSA activities, since several of the adaptation practices discussed in this book also lead to carbon storage in soils and biomass.
- Secure sustained long-term budget allocations for extension services and evaluate the possibility of matching budget schemes with the states, which own all land in Nigeria and are responsible for its degradation or productivity.

Water Sector

The Federal Ministries of Power and Water Resources could on a pilot basis use robust decision making or similar techniques in the feasibility studies for specific irrigation and hydropower projects, to ensure that their design is optimized by taking into account a wide range of climate change scenarios. At the design stage, climate change impact should be quantified and the infrastructure made more robust. The methodology used in this book provides examples of application of robust decision making, which should be part of a broader basin-scale approach that addresses issues like environmental flow, other user demands, sedimentation, evaporation, etc.

Furthermore, the Federal Government (Ministries of Water, Agriculture, and Environment), in collaboration with state governments, could develop in a few states integrated watershed management and monitoring plans (accelerating current efforts such as those supported by the EU, the Japan International Cooperation Agency, and the World Bank in the NEWMAP project) to better incorporate climate resilience into watershed management.

Finally, the Federal Government could consider scaling up investments into the hydrometeorological system. NEWMAP has earmarked funds for the agencies related to the Ministry of Water Resources to modernize the system in selected basins. This effort will need expansion and further capacity-strengthening. Given the critical importance of data to managing climate risks, additional financing could come from the Ecological Fund Office.

Detailed Analysis of Climate Risk to Lagos

To complement this assessment of climate risk to Nigeria's agriculture and water sectors, the climate impacts on the urban and coastal sectors should be analyzed in detail, with emphasis on the Lagos metropolitan area.

The Lagos State Government has already embarked on a series of climate-related activities, largely through the State Ministry of Environment. It organized high-profile international summits on climate change in 2009 and 2011. The Ministry of Environment has embarked on a concerted program of awareness activities, focused on schools and active engagement of young people; an urban greening program is underway; and the ministry is setting up a weather unit that would coordinate with NIMET and other climate information centers to lay the foundation for an early warning system. The Lagos State Emergency Management Agency was established in 1997, and there is an active program of activities to strengthen emergency coordination and response, including designation of emergency relief camp locations and setting up volunteer community emergency response groups.

These initiatives should be augmented by detailed economic analysis of the climate risks in Lagos and identification and prioritization of adaptation measures. The three main areas to be assessed are climate change influence on

- Sea level, river flows, and coastal inundation in the Lagos metropolitan area
- Rainfall and local flooding of built-up areas
- Water supply to Lagos.

The assessment should estimate the costs of climate change, taking into account the value of future land use lost due to more frequent climatic extremes.

A prerequisite for such climate risk analysis is detailed topographical data in digital format. A digital terrain model for the Lagos metropolitan area and bathymetric data of the coasts will make it possible to estimate economic value and damage costs in the detail necessary for an urban environment.

References

IFPRI (International Food Policy Research Institute). 2010. "Strengthening Innovation Capacity of Nigerian Agricultural Research Organizations." IFPRI Discussion Paper 01050, Washington, DC.

JICA (Japan International Cooperation Agency). 2012. *The Project for Review and Update of Nigeria National Water Resources Master Plan.* Report prepared for the Federal Ministry of Water Resources, Abuja, Nigeria.

Liniger, H. P., R. Mekdaschi Studer, C. Hauert, and M. Gurtner. 2011. *Sustainable Land Management in Practice—Guidelines and Best Practices for Sub-Saharan Africa.* TerrAfrica, World Overview of Conservation Approaches and Technologies (WOCAT), and Food and Agriculture Organization of the United Nations (FAO).

World Bank. 2008. "Nigeria: Agriculture Public Expenditure Review." Report No. 44000-NG, Washington, DC.

World Resources Institute (WRI), in collaboration with United Nations Development Programme, United Nations Environment Programme, and World Bank. 2011. *World Resources 2010–2011: Decision Making in a Changing Climate—Adaptation Challenges and Choices.* Washington, DC: WRI.

APPENDIX A

Spatial Disaggregation

This appendix provides an overview of the different spatial scales used in the analysis of climate change impacts and adaptation options. These scales are illustrated in map A.1.

Crop analysis was based on agro-ecological subzones (AESZ) that were hand-digitized in GIS from http://soilsnigeria.net/Publications/Others/new_agroeco_map.jpg, an updated map specific for Nigeria developed by the National Special Programme for Food Security (NSPFS). There is more spatial detail on known agro-ecological zoning in terms of climate and land morphology characteristics, so it is useful for better distinction and discussion of land management for agriculture at a broader spatial level.

Water resources were analyzed at subwatershed level to match the spatial dynamics of hydrological processes. After simulating the river network through the digital elevation model, 893 sub-basins were automatically extracted and each was associated with one of the eight hydrological areas (HA) that is the basis for planning and water management by the federal government and River Basement Development Authorities (RBDAs). The HA map was hand-digitized through GIS tools from http://www.fao.org/DOCREP/005/T1230E/T1230E02.htm.

Economic analysis was performed on agro-ecological zones (AEZs) as an aggregation of AESZ (see table 3.4 in the main text). The AEZ approach used in the computable general equilibrium (CGE) was a first step toward improving description of land use patterns. Following Food and Agriculture Organization of the United Nations (FAO) and International Institute for Applied Systems Analysis (IIASA) methodology, the world land endowment is split into 18 AEZs (https://www.gtap.agecon.purdue.edu/resources/res_display.asp?RecordID=1900), of which Nigeria comprises six. The AEZ database for economic modeling identifies crops and forest extent and production for each region by AEZ. By embedding the AEZ approach into CGE, land is now assumed to be suitable to different uses within, but not between, AEZs. This implies, for instance, that in a given country crop-switching is

Map A.1 Scales of Spatial Disaggregation Used in the Analysis

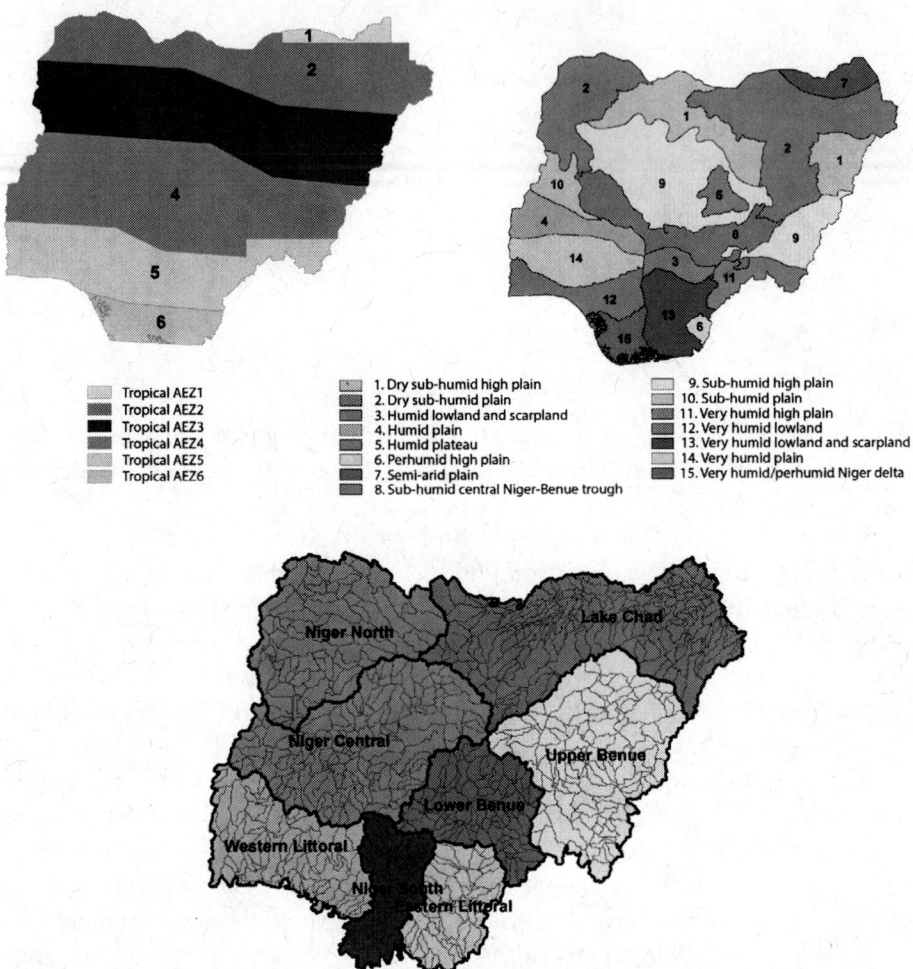

Note: The maps show the different levels of disaggregation of the Nigerian territory used in the analysis: agro-ecological zones (top left); agro-ecological subzones (top right); and eight hydrological areas and internal sub-basins (bottom).

possible only for crops within the same AEZ and that land elasticity of transformation in principle could differ between AEZs. Land substitution mechanisms are thus much more realistic than not using AEZs; they better capture the biological and geographical characteristics of different types of land.

APPENDIX B

Climate Analysis: Current Variability and Future Change

Historical and Current Climate Variability

Climate simulations were performed covering 1970–2065 from a global Coupled General Circulation Model (CGCM), CMCC-MED (about 80 km of horizontal resolution; Scoccimarro *et al.* 2011). The six-hourly CGCM outputs were used as boundary conditions to run a regional climate model (RCM), COSMO Model in Climate Mode (COSMO-CLM) (about 8 km of horizontal resolution; Rockel, Will, and Hense 2008) for Nigeria. The RCM outputs are now available for 1970–2065 at daily frequency.

The regional model capability to represent the current climate is here discussed comparing observed two-meter air temperature (t2m hereafter) and surface precipitation from (gridded) interpolated Climate Research Unit (CRU) observations with the RCM results for the historical period, 1976–2005.

In terms of t2m, the regional model does a good job of reproducing the positive trend observed by CRU during 1976–2005. The simulated t2m averaged for Nigeria for 1976–2005 is 27.2°C, whereas the mean value from CRU interpolated observations for the same period is 27.0°C. The model also reproduces a realistic seasonal variation of t2m averaged for Nigeria. The model does a good job of keeping April and October observed maxima despite temperature overestimation from January to September and underestimation from October to December.

Moreover, both model results and observations show an increase in 1996:2005 t2m seasonal cycle compared with what was observed for 1976–85.

Averaged precipitation for Nigeria for 1976–2005 from CRU interpolated observation is about 3.1 mm/day and there are no significant trends in the annual averaged time series in the present climate period. Also, the precipitation modeled does not show any significant trend and has a bias of about 0.3 mm/day, resulting in an averaged value of 3.4 mm/day for the entire period. The model's positive bias is mainly due to an overestimation for autumn with a maximum of 1 mm/day in October. The summer precipitation patterns are smoothed in the observations compared to the model patterns where orographic features are more evident. Also for Ondo, Edo, and Delta states the model shows an evident positive bias for summer. In autumn the modeled precipitation over the Jos plateau and Mambilla mountains persists even though in the observations the precipitation signal associated to the orography is lost during this season. In winter precipitation is nearly absent at latitudes higher than 8°N both in observations and the model. For winter the model tends to underestimate precipitation. This effect, associated to the overestimation found in autumn, induces a more pronounced seasonal cycle in the model than the one observed.

The good performances of the model were also confirmed by a site comparison. Indeed, modeled precipitation and t2m were further validated using meteorological data for 20 of the Nigerian Meteorological Agency (NIMET) stations, the ones particularly suitable in terms of length and continuity of time series (thus avoiding large inter- and intra-annual gaps) and spatial representativeness (different elevations and subregions). From this analysis, useful data from 1990 to 2009 were valuable for comparing temperature and from 1985 to 2005 for comparing precipitation. However, as only maximum and minimum temperature data were available from observations, the average was used as the best t2m approximation for comparison with t2m from the model.

Bias Correction for the Present Climate

According to the recent work of Sperna *et al.* (2010), for the bias-correction of RCM outputs monthly scaling factors were calculated from the difference (temperature) or ratio (precipitation) of the 30-year average monthly means (1976–2005) between the observation dataset and regional simulation outputs.

For temperature, an additive correction is used:

$$T_{Rcorr} = T_R + (\dot{T}_{obs} - \dot{T}_R) \tag{B.1}$$

where T is the daily temperature and $\dot{T}$ is the 30-year average monthly temperature. Subscript 'R' indicates 'original' regional outputs and 'Rcorr' the corrected one; 'obs' stands for observed values.

For precipitation a multiplicative correction is used:

$$P_{Rcorr} = P_R \times (\dot{P}_{obs} / \dot{P}_R) \tag{B.2}$$

where P is the daily precipitation and $\dot{P}$ is the 30-year average monthly precipitation.

As differences between the modeled and observed monthly precipitation amount and number of wet days can be very large, a simple multiplicative correction could result in unrealistic precipitation peaks in the bias-corrected precipitation time series. Therefore the bias-correction of the precipitation equation is extended with a minimum daily precipitation amount that has to be exceeded by the modeled total monthly precipitation amount before the multiplicative correction can be used. This threshold (P_{thr}) is

$$P_{thr} = (\dot{P}_{obs} / \dot{W}_{obs}) \qquad (B.3)$$

where $\dot{W}_{obs}$ is the 30-year average number of wet days for the specific month. In addition, a threshold value of 10 is set for the maximum value of the multiplicative correction factor in equation B.2. In practice, if (a) the monthly sum of regional precipitation does not exceed the threshold P_{thr} or if (b) the multiplicative correction factor is higher than 10, the days when precipitation occurred are calculated from a temperature limit below which a day becomes wet. With this method the number of wet days is increased to limit large rain events on the few days with rain in the RCM time series.

Perturbation Methodology for Future Climate

The approach to perturb RCM outputs using the variability of global simulations is here described (Buishand and Lenderink 2004).

To take into account both the temporal and the spatial variability of global simulations for Nigeria, for each global simulation g, for each grid point c, and for each time step t (day) in the regional simulation, perturbed high-resolution precipitations (P_{Rp}) are calculated as

$$P_{Rp}(g,c,t) = P_R(c,t) \times (\dot{P}_G(g,c) / \dot{P}_R(c)) \qquad (B.4)$$

while perturbed high-resolution temperatures (T_{Rp}) are calculated as

$$T_{Rp}(g,c,t) = T_R(c,t) + (\dot{T}_G(g,c) - \dot{T}_R(c)) \qquad (B.5)$$

where, on the right-hand side:

P_R and T_R are the values of the regional precipitation and temperature grids.

$\dot{P}_G$ and $\dot{T}_G$ represent the monthly global model averages (for the month including the day t) of precipitation and temperature from the global model.

$\dot{P}_R$ and $\dot{T}_R$ represent the monthly regional model averages (for the month including the day t) of precipitation and temperature.

References

Buishand, T. A., and G. Lenderink. 2004. *Estimation of Future Discharges of the River Rhine in the SWURVE Project.* KNMI Technical Report, Koninklijk Nederlands Meteorologisch Instituut, De Bilt, Netherlands.

Rockel, B., A. Will, and A. Hense. 2008. "The Regional Climate Model COSMO-CLM (CCLM)." *Meteorologische Zeitschrift* 17 (4): 347–48.

Scoccimarro, E., S. Gualdi, A. Bellucci, A. Sanna, P. G. Fogli, E. Manzini, M. Vichi, P. Oddo, and A. Navarra. 2011. "Effects of Tropical Cyclones on Ocean Heat Transport in a High-Resolution Coupled General Circulation Model." *Journal of Climate* 24: 4368–84.

Sperna, W. F. C., L. P. H. van Beek, J. C. J. Kwadijk, and M. F. P. Bierkens. 2010. "The Ability of a GCM-Forced Hydrological Model to Reproduce Global Discharge Variability." *Hydrological Earth System Sciences* 14: 1595–1621. doi:10.5194/hess-14-1595-2010.

APPENDIX C

Crop Modeling

Method of Analysis

The CSM-DSSAT-CSM (Decision Support System for Agrotechnology Transfer—Cropping System Model), introduced in chapter 3, is structured using a modular approach described by Jones, Keating, and Porter (2001) and Porter, Jones, and Braga (2000). The DSSAT-CSM has models for sorghum, rice, millet, maize, and cassava, but not for yams.

Figure C.1 shows the DSSAT-CSM scheme, with the main tools and crop simulation models implemented.

The DSSAT-CSM simulates the growth, development, and yield of a crop growing on a uniform area of land under prescribed or simulated management and the changes in soil water, carbon, and nitrogen that take place in a cropping system over time.

Databases describe weather, soil, experimental conditions and measurements, and genotype information for applying the models to different situations (Jones *et al.* 2003). There are also improved application programs for seasonal and sequence analyses that assess the environmental impacts of irrigation management, climate change and variability, and precision management. Moreover, it is possible to change the ambient CO_2 concentration, which is very important in climate change impact studies because it has effects (in particular for C3 crops) on photosynthesis (biomass accumulation) and water use efficiency (considering stomatal conductance).

DSSAT-CSM software defines "minimum data set" (MDS), the minimum required to run the crop models and validate the outputs (International Consortium for Agricultural System Applications, http://www.icasa.net/dssat /minimum.html). The MDS includes:

- Site weather data for the entire growing season, such as
  - Latitude and longitude of the weather station
  - Daily values of incoming solar radiation (MJ/m²-day)
  - Maximum and minimum air temperature (°C)
  - Rainfall (mm).

Figure C.1 DSSAT-CSM Scheme

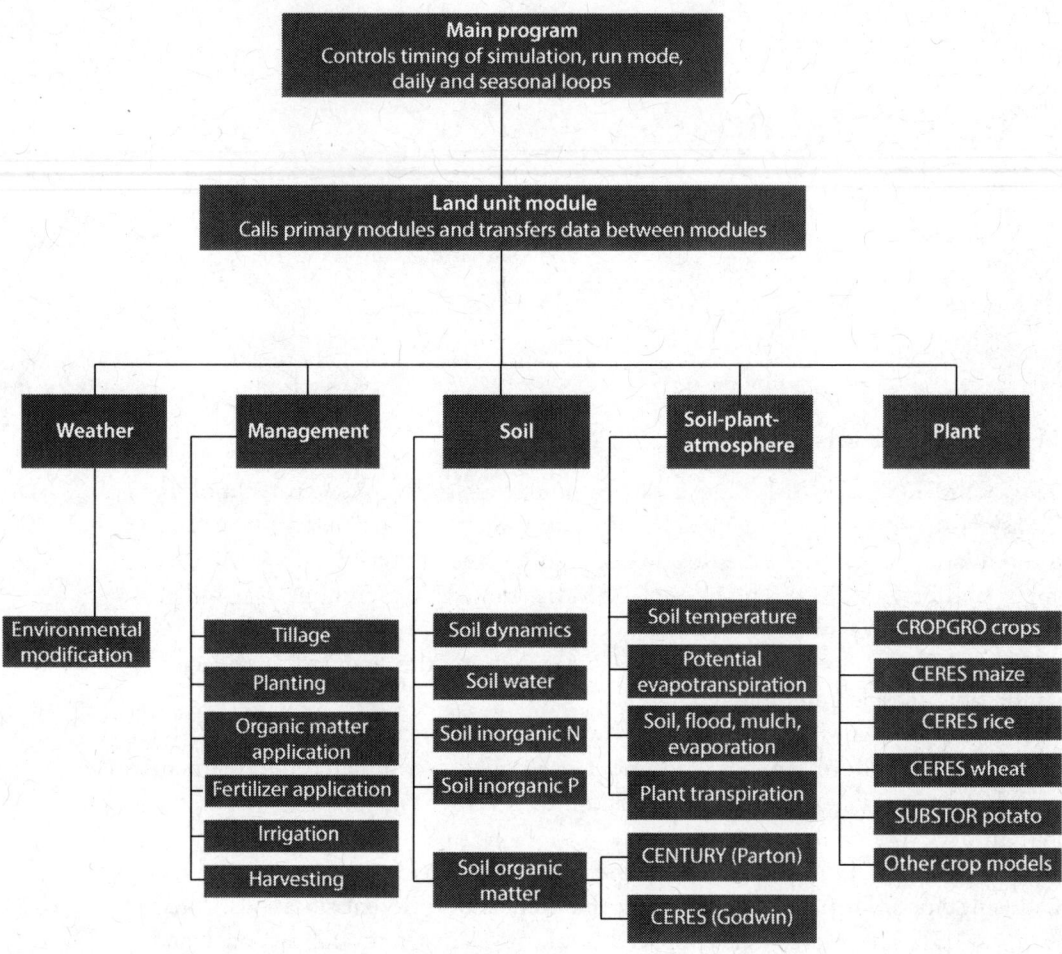

Source: Modified from Porter *et al.* 2009.

- Site soil data: soil classification (e.g., USDA/NRCS), surface slope, color, permeability, and drainage class. Data by soil horizons include upper and lower horizon depths (cm); percentages of sand, silt, and clay; one-third bar bulk density; organic carbon; and pH in water.
- Management and observed data from an experiment: information on planting date, dates when soil conditions were measured before planting, planting density, row spacing, planting depth, crop variety, irrigation, and fertilizer practices.

Methodology

The methodology for crop modeling analysis includes (a) model set-up (collection of the MDS required by the models; preparation of crop, soil, and weather databases; adapting cassava model coefficients to simulate the yam crop); (b) crop simulation models for calibration and validation; and (c) assessment of

Figure C.2 Scheme of Methodology

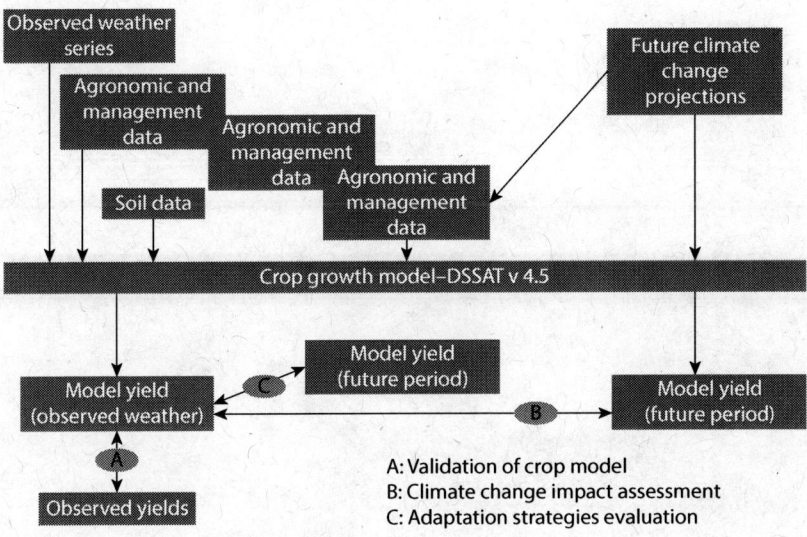

A: Validation of crop model
B: Climate change impact assessment
C: Adaptation strategies evaluation

Source: Dubrovsky 2009, modified.

impact of projected climate change conditions. Finally, adaptation strategies were suggested to cope with the projected climate change conditions (figure C.2).

Numerous combinations of soil and climate conditions were considered for each Nigerian agro-ecological zone, in which specific crop management and varieties were set based primarily on USAID MARKET 2010.

Different agro-ecological subzones (AESZs; map C.1) were identified, and for each AESZ different crop management options, such as growing periods and/or crop varieties cultivated (long or medium growing season), based on USAID MARKETS (2009a, 2009b, 2010), were considered. The other management options were the same for each run: no irrigation and the same mechanization, fertilization regime, plant density, and so forth.

For each crop, only the AESZs where the crops are most diffused were considered in the aggregation for the entire country. In particular, for sorghum the AESZs were 1, 2, 8, 9, and 11; for maize AESZs 3, 4, 5, 8, 9, 11, 12, 13, 14, and 15; for millet AESZs 1, 2, 4, 5, 8, 9, 11, and 14; for rice AESZs 1, 2, 3, 4, 8, 9, 10, 11, 12, 13, 14, and 15; for cassava AESZs 3, 4, 5, 8, 9, 11, 12, 13, 14, and 15; and for yams AESZs 3, 4, 5, 8, 9, 11, 12, 13, 14, and 15.

Yam crop modeling required special attention. Because the DSSAT software does not have a simulation module for yams, the objective was to identify a set of genetic coefficients for simulating yam growth and yield. Cassava crop parameters (included into DSSAT) were used as a starting point to set the parameters for yams, as in previous studies (Srivastava and Gaiser 2010).

Climate impact was assessed by comparing the yield obtained with the weather data for the present period and the yield projected to be obtainable

Map C.1 Agro-ecological Subzones of Nigeria

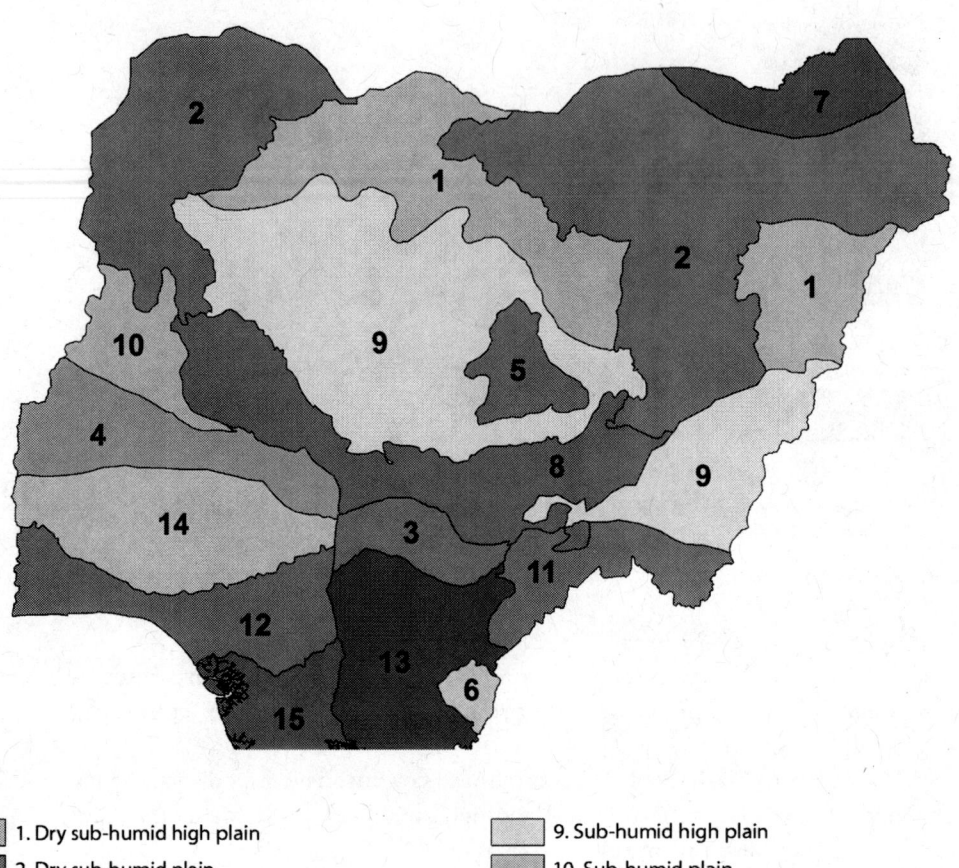

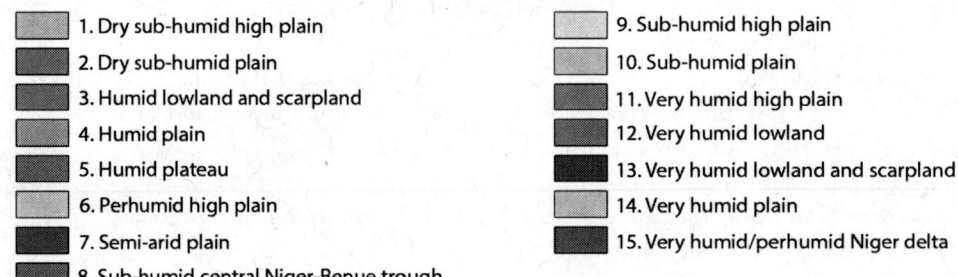

1. Dry sub-humid high plain		9. Sub-humid high plain	
2. Dry sub-humid plain		10. Sub-humid plain	
3. Humid lowland and scarpland		11. Very humid high plain	
4. Humid plain		12. Very humid lowland	
5. Humid plateau		13. Very humid lowland and scarpland	
6. Perhumid high plain		14. Very humid plain	
7. Semi-arid plain		15. Very humid/perhumid Niger delta	
8. Sub-humid central Niger-Benue trough			

under changed climate conditions. The model runs were made keeping fixed all the other input parameters (e.g., crop management, soil, etc.) and varying only the weather data.

Data from the unperturbed regional climate model (RCM) and five of its perturbations were used in impact analysis. The climate projections derived by CMCC-MED, Geophysical Fluid Dynamics Laboratory (GFDL), Institute of Atmospheric Physics (IAP), MIROC, and National Center for Atmospheric Research (NCAR) were selected to explore the range of uncertainties in climate projections because they project the most extreme changes in temperature and precipitation.

Climate change impacts were calculated for the baseline period 1976–2005 (centered on 1990) and for the 30-year climatic periods centered on 2020 (2006–35) and 2050 (2036–65).

In this study, direct and indirect effects of different CO_2 concentrations, projected for the future periods, were explored separately to estimate the effect linked to it. Crop models were first run based on a fixed value of CO_2 concentration (380 ppm) to explore the indirect effect of CO_2 concentration due to changed climate conditions. Then models were run to consider future projected increases in CO_2 atmosphere concentration (from 380 to 582, under A1B), to look at combined direct and indirect effects.

Increased CO_2 in the atmosphere has a direct effect on crops, and it also has a fertilization effect because of a higher photosynthesis rate (in particular for C3 crops), but CO_2 could also affect the transpiration rate, improving water use efficiency because of a modification in stomatal conductance when there is a high concentration of CO_2.

Although the model calibration/validation is made for only a few Nigerian states, the climate impact is assessed for all Nigerian areas where each crop is grown. The higher resolution of the RCM makes this possible. Data were aggregated into different AESZs and expressed as percentage of yield change for 2020 and 2050 with respect to the baseline.

A set of adaptation options was then singled out to analyze their potential to offset, across space (the different AESZs), time (2020 and 2050), and crops, the negative impacts of climate change on yields. The selection was dictated by data availability and by the suitability of integrating the option chosen into the crop model used to evaluate climate impacts. For rain-fed areas, seven adaptation options (table 6.1) were analyzed clustered as follows:

- Shift of the sowing/planting date (one month earlier or later than the traditional calendar)
- Conservation/organic agriculture practices, including management of manure and residues
- Use of inorganic fertilizers.

For irrigated crops, the analysis focused on yield improvements that could be achieved by modifying planting/sowing dates. To address model uncertainty, climate data from the RCM model and the two extreme perturbations were considered. CO_2 atmospheric concentration was kept constant.

Results

Results of crop simulations for sorghum, millet, maize, rice, cassava, and yam yields after climate change are reported here as

- Maps of crop yield changes for sorghum, maize, millet, rice, cassava, and yams (percentage in each AESZ), considering the COSMO Model in Climate Mode

(COSMO-CLM) RCM separately, with or without considering increasing atmospheric CO_2

- Graphs with crop yield changes (percentage in each AESZ) considering the COSMO-CLM RCM and five COSMO-CLM perturbations separately, with or without considering increasing atmospheric CO_2 concentration.

Results of crop simulations of crop yields after climate change, including adaptation strategies, are reported here as tables of yield changes with respect to the baseline for each crop, in each AESZ, with RCM and the two extremes perturbed models (GFDL and NCAR).

Figures C.3 and C.4 show the impacts on crop yields for rice for 2020 and 2050 without and with considering increasing atmospheric CO_2 values for unperturbed RCM and five perturbations. Similar graphs for crop yield changes are also available for maize, millet, sorghum, and yams but are not shown here.

Keeping the CO_2 concentration fixed, yield reduction can be expected for all crops, with some differences related to AESZ and climate model. However, this negative effect of changed climate conditions is partially reduced by the increased atmospheric CO_2 concentration, proving that the direct effect of CO_2 concentration is to partially reduce the negative impact due to changed weather conditions.

The climate change impacts on crop yields obtained by perturbing the RCM COSMO-CLM model with data from GCM models confirm the negative effects of changed climate conditions, especially for sorghum, millet, and rice, and, conversely, the positive effects of direct CO_2 on crop production, which may dominate the negative effect of changed weather conditions.

The key messages of this study are that reductions in yield are recorded for all crops in both 2020 and particularly in 2050, although there are differences between crops and AESZs. These reductions are mainly due to higher temperature that shortens the crop growing cycle and consequently lowers biomass accumulation, but they are affected by changes in precipitation patterns. These negative effects due to changed climate conditions seem to be partially, and in some cases totally, mitigated by increases in atmospheric CO_2 concentration. The positive effect is particularly evident in C3 crops, whose physiology better responds to higher CO_2, improving photosynthesis rates. Another important effect, for both C3 and C4 species, is stomata closure, which improves water use efficiency especially in areas with low precipitation, like the northern AESZs.

Table C.1 demonstrates for rice the climate change impacts on crop yields obtained considering several adaptation options. The results reported for each AESZ and period (2020 and 2050) are average values of projections obtained with COSMO-CLM RCM and its perturbations with the two extremes GCM models (GFDL and NCAR). Similar tables are available from the study team for the other crops: sorghum, maize, millet, and yams.

Figure C.3 Impacts on Rice Yield for 2020 and 2050 by AESZ; CO_2 Concentration of 380 ppm; with RCM and Its Perturbations

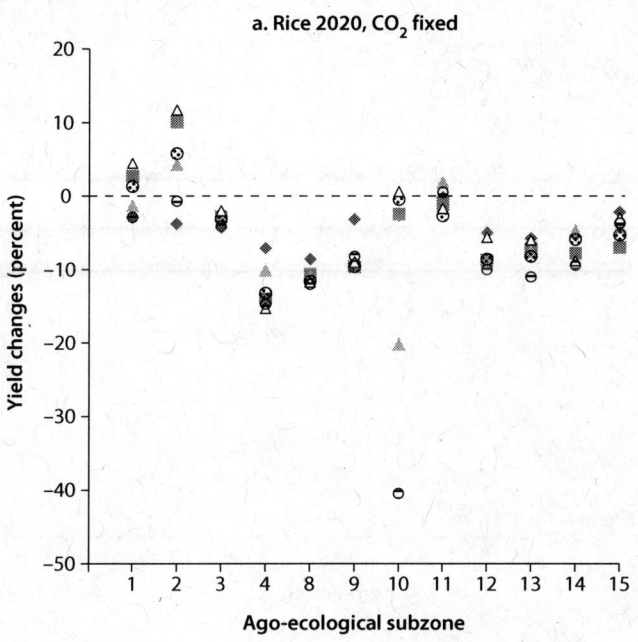

a. Rice 2020, CO_2 fixed

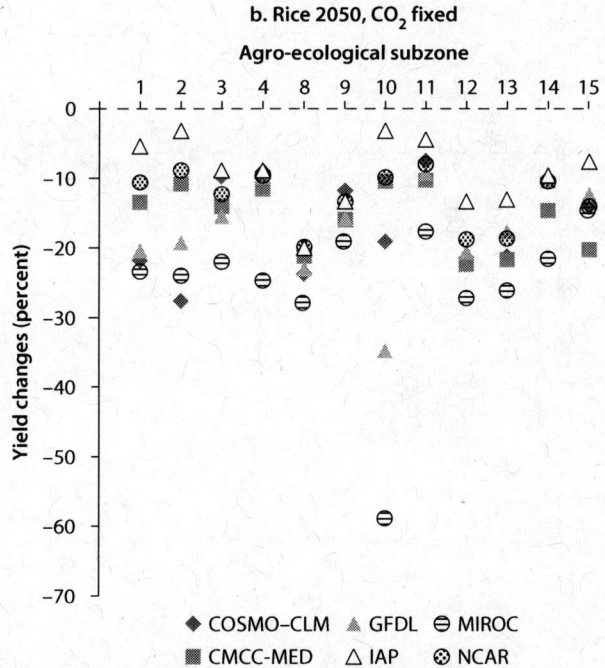

b. Rice 2050, CO_2 fixed

Source: Authors' calculations based on data sources listed in table 3A.1.
Note: COSMO-CLM = COSMO Model in Climate Mode; CMCC-MED = Euro-Mediterranean Center on Climate Change; GFDL = Geophysical Fluid Dynamics Laboratory; IAP = Institute of Atmospheric Physics; MIROC = Center for Climate System Research; NCAR = National Center for Atmospheric Research; RCM = Regional Climate Model.

Figure C.4 Impacts on Rice Yield for 2020 and 2050 by AESZ Changing CO_2 Concentration

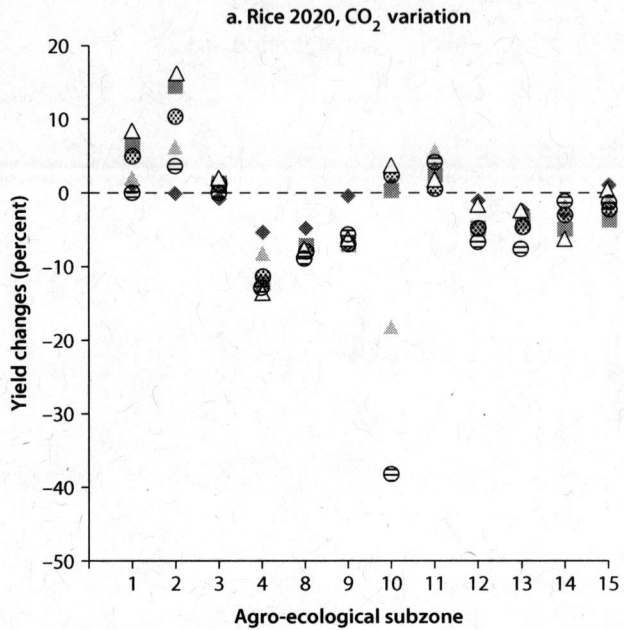

a. Rice 2020, CO_2 variation

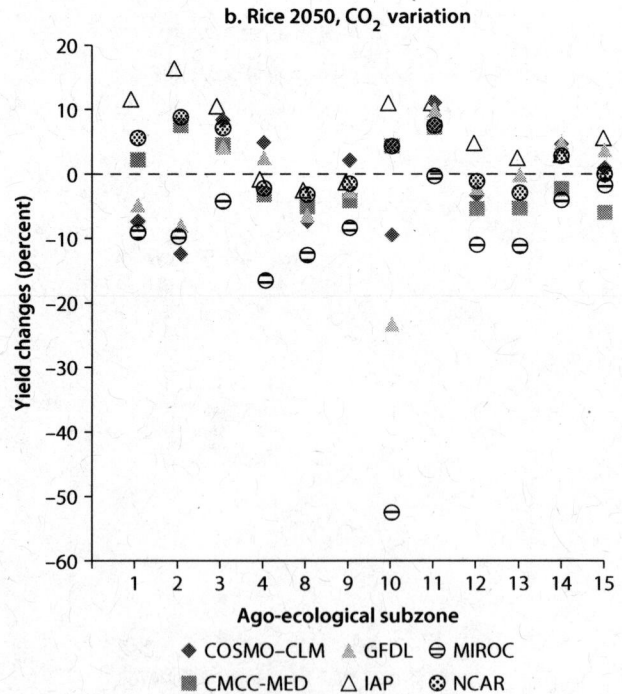

b. Rice 2050, CO_2 variation

◆ COSMO–CLM   ▲ GFDL   ⊖ MIROC
■ CMCC-MED   △ IAP   ⊗ NCAR

Source: Authors' calculations based on data sources listed in table 3A.1.
Note: COSMO-CLM = COSMO Model in Climate Mode; CMCC-MED = Euro-Mediterranean Center on Climate Change; GFDL = Geophysical Fluid Dynamics Laboratory; IAP = Institute of Atmospheric Physics; MIROC = Center for Climate System Research; NCAR = National Center for Atmospheric Research.

Table C.1 Maximum and Minimum Climate Change Impacts on Rice Yield by Adaptation Strategy

	AESZ	−1 month rain-fed		+1 month rain-fed		Fertilizer 1 rain-fed		Fertilizer 2 rain-fed		Manure 1 rain-fed		Manure 2 rain-fed		Residuals 1 rain-fed		−1 month irrigated		+1 month irrigated		Irrigated	
2020	1	-1.6	4.8	-26.0	-16.8	0.7	6.0	1.7	11.2	1.1	11.2	-23.4	-11.5	1.0	11.1	-4.1	3.8	-12.6	-2.1	0.7	7.3
	2	0.1	7.6	-22.2	-5.5	-1.4	7.8	-0.6	12.6	0.1	14.7	-10.2	-1.8	0.0	16.7	2.4	9.7	8.7	17.4	12.1	16.9
	3	-3.8	5.0	-18.6	-5.2	-1.1	3.7	-0.6	5.3	0.3	5.3	-21.3	-9.7	0.2	5.3	-4.8	4.4	-21.3	-9.8	-4.3	-2.6
	4	-5.5	1.4	-26.5	-17.6	-0.4	1.8	2.0	7.9	3.8	10.3	-14.6	2.3	3.3	9.8	-6.3	-2.0	-28.0	-17.9	-13.2	-7.1
	8	-16.7	-14.0	-13.8	-10.9	-9.5	-7.1	-8.4	-5.1	-7.5	-3.6	-17.4	-10.6	-7.6	-3.7	-19.3	-17.3	-17.5	-13.3	-11.6	-7.9
	9	-6.8	-1.2	-20.3	-17.2	-3.3	1.5	-1.8	2.8	0.5	4.1	-28.3	-7.3	0.3	4.0	-7.9	-3.1	-21.8	-11.9	-9.6	-2.5
	10	-26.6	7.6	-43.5	-28.6	-12.8	8.4	-9.9	15.1	5.5	28.9	-35.9	-16.4	5.4	28.5	-17.6	5.8	-24.0	2.0	-14.4	6.7
	11	1.3	9.2	-22.2	-14.5	7.5	8.0	8.5	14.2	8.2	11.1	-33.5	-12.0	8.1	10.9	1.0	4.6	-25.1	-15.3	-2.7	2.1
	12	-4.2	1.9	-24.9	-20.4	-4.4	-1.3	-3.1	1.5	-1.1	2.3	-17.2	-2.7	-1.3	2.2	-6.3	1.2	-27.1	-23.5	-3.5	-0.2
	13	-10.0	0.9	-29.0	-22.2	-1.2	1.2	0.9	5.0	4.1	6.8	-8.3	1.6	3.9	6.6	-11.8	-5.0	-30.6	-21.9	-6.5	-4.1
	14	-0.2	3.6	-16.3	-13.1	1.5	2.8	4.1	7.7	6.0	8.0	-1.7	1.5	5.8	7.8	-2.0	-0.5	-18.3	-12.6	-5.9	-0.8
	15	2.8	11.1	-26.2	-23.9	5.8	8.5	8.9	13.4	13.6	17.1	-5.8	12.4	13.2	16.7	4.7	10.0	-28.6	-23.8	-5.1	-1.9
2050	1	-25.4	0.4	-38.3	-11.2	-19.4	-7.0	-18.8	12.2	-18.9	12.9	-27.8	-19.7	-19.0	12.8	-28.2	-18.3	-20.1	-16.8	-17.7	-10.3
	2	-32.1	0.5	-39.4	0.6	-26.9	-7.0	-26.7	12.6	-26.0	14.0	-30.0	-6.7	-26.1	16.9	-33.7	-17.2	-7.8	-1.6	-12.8	-4.9
	3	-20.9	2.9	-20.1	1.7	-14.6	-6.5	-14.4	9.5	-13.3	11.8	-19.7	-10.4	-13.3	11.8	-21.1	-10.5	-22.9	-17.9	-15.5	-9.7
	4	-13.4	6.7	-17.2	-12.8	-7.2	-0.6	-7.0	17.4	-6.9	19.7	-7.3	-1.4	-6.9	19.5	-16.0	-9.1	-24.4	-13.3	-9.8	-9.0
	8	-31.9	-12.8	-22.5	-5.7	-23.3	-19.2	-23.3	-1.1	-22.7	0.8	-23.5	-10.9	-22.7	0.7	-35.5	-30.6	-23.6	-23.2	-23.3	-19.9
	9	-18.9	1.7	-19.8	-12.8	-13.0	-8.6	-12.1	8.1	-10.5	9.2	-22.0	-14.1	-10.6	8.9	-20.2	-13.3	-23.2	-15.4	-13.5	-11.1
	10	-40.3	-0.1	-42.0	-31.0	-31.1	-3.1	-29.6	16.1	-10.4	34.0	-40.3	-30.6	-10.6	33.9	-31.4	-16.1	-25.8	-12.7	-21.7	-8.9
	11	-14.3	11.5	-16.6	-14.1	-7.7	-2.6	-7.4	19.1	-6.6	18.9	-28.0	-13.5	-6.6	18.6	-15.3	-8.1	-25.8	-14.5	-9.2	-7.6
	12	-23.1	-1.5	-30.2	-20.3	-19.0	-16.5	-18.6	4.3	-16.9	5.5	-19.6	-11.0	-17.0	5.4	-26.2	-20.8	-32.3	-30.0	-18.3	-13.8
	13	-28.2	-5.2	-30.3	-24.9	-17.1	-14.7	-16.0	6.4	-13.7	8.8	-16.4	-3.3	-13.8	8.6	-31.0	-24.5	-33.8	-26.6	-21.1	-15.4
	14	-10.9	8.7	-15.4	-9.1	-6.5	-5.0	-5.3	15.0	-2.6	16.3	-5.9	7.6	-2.8	16.0	-14.3	-9.3	-19.5	-14.7	-10.5	-7.1
	15	-12.7	9.3	-30.7	-22.2	-7.7	-6.4	-6.1	15.2	-3.7	19.5	-5.2	-0.8	-3.8	19.1	-12.7	-9.7	-33.2	-30.2	-13.7	-11.8

Source: Authors' calculations based on data sources listed in table 3A.1.

131

References

Dubrovský, M. 2009. "Linking the Climate Change Scenarios and Weather Generators with Agroclimatological Models." Paper presented at a seminar in Sassari, Italy, May 25–June 5. http://www.ufa.cas.cz/dub/crop/2009-sassari-martin-seminar-part3.pdf.

Jones, J. W., G. Hoogemboom, C. H. Porter, K. J. Boote, W. D. Batchelor, L. A. Hunt, P. W. Wilkens, U. Singh, A. J. Gijsman, and J. T. Ritchie. 2003. "The DSSAT Cropping System Model." *European Journal of Agronomy* 18: 235–65.

Jones, J.W., B.A. Keating, C. H. Porter. 2001. "Approaches to Modular Model Development." *Agricultural Systems* 70: 421–443.

Porter, C. H., J. W. Jones, S. Adiku, A. J. Gijsman, O. Gargiulo, and J. B. Naab. 2009. "Modeling Organic Carbon and Carbon-Mediated Soil Processes in DSSAT v4.5." *Operational Research* 10 (3): 247–78. doi:10.1007/s12351-009-0059-1.

Porter, C. H., J. W. Jones, and R. Braga. 2000. "An Approach for Modular Crop Model Development." International Consortium for Agricultural Systems Applications, Honolulu, HI, 13. http://icasa.net/modular/index.html.

Srivastava, A. K., T. Gaiser. 2010. "Simulating Biomass Accumulation and Yield of Yam (Dioscorea alata) in the Upper Ouémé Basin (Benin Republic). I. Compilation of Physiological Parameters and Calibration at the Field Scale." *Field Crops Research* 116: 23–29.

USAID MARKETS. 2009a. "Package of Practices for Rice Production." http://www.nigeriamarkets.org/.

———. 2009b. "Package of Practices for Sorghum Production." http://www.nigeriamarkets.org/.

———. 2010. "Package of Practices for Maize Production." http://www.nigeriamarkets.org/.

APPENDIX D

Food Security Analysis

Results on crop yields (discussed in chapter 5) were obtained for a reduced subensemble of climate projections consisting of Regional Climate Model (RCM) simulation and for its two most extreme General/Global Circulation Model (GCM) perturbations in terms of climate change impacts: the National Center for Atmospheric Research (NCAR), for the most optimistic case, and the Geophysical Fluid Dynamics Laboratory (GFDL), for the most pessimistic case. They were then integrated with nutritional outcomes, demographic changes, and market access to quantify Nigeria's future food security threats by agro-ecological subzones (AESZs). The results were used to analyze possible food security threats by calculating the mean adequacy ratio (MAR) for the present and for 2020 and 2050.

MAR measures for the population as a whole the degree to which available food crops fulfill dietary energy and nutrient requirements. It is calculated by averaging individual nutrient adequacy ratios (NAR; Hatløy, Torheim, and Oshaug 1998). The estimated NAR specific to calories or nutrients is defined as the per-person ratio of energy or nutrients available from food crop quantities beyond the recommended nutrient intake (RNI). NAR and MAR both equal 1 when average intake of energy and proteins corresponds to the recommended intake; lower than 1 they identify nutrient deficiency.

For this study, MAR in Nigeria was defined as the average of NAR from calories and proteins. Following World Health Organization (WHO) recommendations, an average caloric intake of 2,380 Kcals per person per day (basal metabolic rate of 1,400 Kcals per person per day) and protein intake of 40 grams per person per day (average weight per person 55 kg) was estimated following WHO indications (table D.1). The recommended nutrient intake is calculated as a function of body weight and basal metabolic rate.

The Spatial Production Allocation Model (SPAM) geodataset (You *et al.* 2010) specifies physical and harvested area, yields, and production in 2000 of main crop typologies. SPAM uses a variety of geographic themes (land cover/land use, biophysical crop suitability, population density, and distance to markets) to spatially redistribute subnational crop production statistics. This analysis used the SPAM dataset to derive spatialized estimates of utilized area, yields, and production in 2000 of the main crops in Nigeria. The equivalent nutritional values by weight (calories and proteins) for these crops are derived from the FAOSTAT 2010[1] database (see table D.2).

The GRUMPv1 dataset (CIESIN *et al.* 2011) spatially defines population density based on statistical interpolation of population census data (for 2000), available for a very high number (~1,000,000) of administrative units, redistributed through nighttime satellite imagery of the intensity distribution of artificial light (map D.1).

Population growth in Nigeria is currently quite high; it averaged 2.5–2.7 percent annually between 1970 and 2010. In 2000 the population was already 120 million people, and by 2010, at a steady growth rate of 2.6 percent, it had surpassed 150 million. The UN projects that by 2050 there will be 390 million

Table D.1 Recommended Nutrient Intake of Calories and Protein per Person

	Recommended nutrient intake
Energy (Kcals per person per day)	Male: 1.78 × basal metabolic rate Female: 1.64 × basal metabolic rate
Protein (grams per person per day)	Male & female: body weight × 0.75 g

Source: WHO 1985.

Table D.2 Nutritional Values of Nigeria's Main Crops

	Calories (Kcals per 100 grams)	*Protein (mg per gram)*
Cassava	109	9
Maize	356	95
Millet	340	97
Rice	280	60
Sorghum	343	101
Yams	101	13

Source: http://faostat.fao.org/.

Map D.1 Population Density and Urban-Rural Extent in Nigeria, 2000

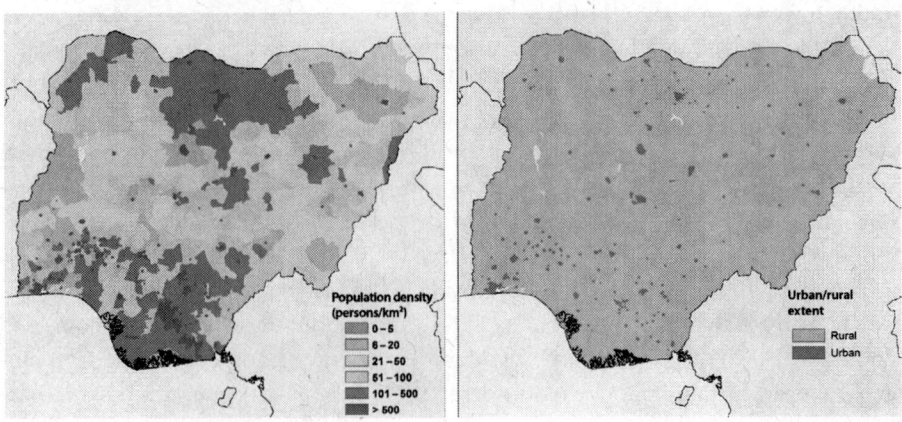

Source: CIESIN *et al.* 2011.

Nigerians. Demographic projections predict average growth in Nigeria of 2.6 percent for 2010–19; 2.5 percent for 2020–29; 2.3 percent for 2030–39; and 2 percent for 2040–49.

At present, almost half of Nigerians are urban dwellers. While average annual population growth for 1980–2000 was 2.7 percent, it varied from 1.7 percent in rural areas to 4.7 percent in urban areas (http://earthtrends.wri .org/). It is projected that through 2050 the growth rate for rural areas will be 1.2 percent and for urban areas 3.8 percent. Population growth distribution has been reconstructed for each AESZ by applying the decade-by-decade growth rates through 2050 for the urban and rural population. In distinguishing between rural and urban area, in addition to simulating differences in birth and mortality rates, the growth rates for each AESZ also simulate migration from rural to urban areas.

Production in 2000 of each major crop type stratified by AESZ was extracted from the SPAM dataset and associated to yield and harvested areas specific for each AESZ. Production may change as a consequence of variations of harvested areas and yields after changes in climate conditions, agronomic practices, or both. As a baseline, changes were assumed only in yields after climate change, which were previously calculated. Specification of impact from these changes alone defines the extent to which either cropland must be expanded or agriculture mechanized to mitigate threats to food security.

Following this baseline approach, total production for each major crop by AESZ was calculated for 2020 and 2050 by integrating the climate change impact on yield (see chapter 5) over current production. Current and future crop production within each AESZ, and for the whole of Nigeria, should satisfy the nutritional requirement of the current and future population. The recommended nutritional intake of an AESZ and the whole of Nigeria is equal to the average recommended daily intake of calories and proteins per person (2,380 Kcal and 40 grams of protein per person) times the total population.

The results of the calculation of NARs for 2000, 2020, and 2050 under RCM are reported in tables D.3, D.4, and D.5, respectively.

Travel time to market centers is used as a proxy for market accessibility and shows the likely extent to which farming households are physically integrated with or isolated from markets. From travel time maps (Nelson 2008), the analysis extracted accessibility using a cost-distance function to measure the cost in time (hours) to the nearest market from each pixel. The friction or adjusted speed varies based on such features as road locations, road type, elevation, slope, country boundaries, water bodies, and land cover (Nelson 2008).

The values of MAR changes by AESZ can then be associated with the average socioeconomic status in each AESZ. The climate change impact on yields of the main crops (see chapter 5) are integrated with nutritional outcomes, demographic changes, and market access to quantify future food security threats by AESZ.

Table D.3 NAR per AESZ, 2000

AESZ	Cassava NAR Kcal	Cassava NAR Prot.	Maize NAR Kcal	Maize NAR Prot.	Millet NAR Kcal	Millet NAR Prot.	Rice NAR Kcal	Rice NAR Prot.	Sorghum NAR Kcal	Sorghum NAR Prot.	Yam NAR Kcal	Yam NAR Prot.	All 6 crops NAR Kcal	All 6 crops NAR Prot.
1	0.00	0.00	0.08	0.13	0.33	0.57	0.04	0.06	0.41	0.72	0.05	0.04	0.93	1.51
2	0.08	0.04	0.06	0.09	0.66	1.13	0.10	0.12	0.48	0.83	0.17	0.13	1.55	2.34
3	1.22	0.60	0.33	0.53	0.00	0.00	0.05	0.07	0.23	0.40	0.42	0.32	2.25	1.91
4	0.69	0.34	0.48	0.76	0.04	0.07	0.10	0.13	0.24	0.41	0.32	0.24	1.86	1.94
5	0.10	0.05	0.14	0.22	0.11	0.18	0.00	0.00	0.31	0.55	0.09	0.07	0.75	1.07
6	0.23	0.11	0.00	0.00	0.00	0.00	0.00	0.00	0.00	0.00	0.54	0.42	0.78	0.53
7	0.00	0.00	0.19	0.30	0.93	1.59	0.00	0.00	2.43	4.26	0.00	0.00	3.55	6.14
8	0.93	0.46	0.34	0.55	0.19	0.31	0.60	0.76	0.70	1.23	1.18	0.90	3.94	4.21
9	0.50	0.25	0.34	0.55	0.27	0.46	0.21	0.26	0.48	0.84	0.49	0.38	2.29	2.73
10	0.39	0.19	0.70	1.11	0.43	0.72	0.47	0.60	1.43	2.51	0.98	0.75	4.40	5.88
11	2.71	1.33	0.33	0.52	0.10	0.17	0.34	0.43	0.39	0.68	1.47	1.13	5.34	4.27
12	0.21	0.10	0.09	0.14	0.00	0.00	0.03	0.04	0.00	0.00	0.11	0.08	0.44	0.36
13	0.56	0.27	0.09	0.15	0.00	0.00	0.03	0.04	0.00	0.00	0.47	0.36	1.15	0.81
14	0.25	0.12	0.12	0.19	0.00	0.00	0.02	0.03	0.00	0.00	0.10	0.07	0.50	0.42
15	0.40	0.20	0.08	0.13	0.00	0.00	0.01	0.02	0.00	0.00	0.18	0.14	0.68	0.48
Total	0.35	0.17	0.14	0.23	0.20	0.34	0.09	0.11	0.25	0.43	0.28	0.22	1.31	1.50

NAR Kcal = Tot_production (Tons) * (nutritional value (Kcal per 100 grams) *10 *1,000)/(population * (2,380 * 365)).
NAR proteins = Tot_production (Tons) * (nutritional value (grams per kg) *1,000)/(population * (40 * 365)).

	Calories (Kcal per 100 grams)	Proteins (mg per gram)
Cassava	109	9
Maize	356	95
Millet	340	97
Rice	280	60
Sorghum	343	101
Yam	101	13

Average caloric recommended intake
2,380 Kcal per day per person

Average protein recommended intake
40 grams per day per person

Source: Authors' calculations based on data sources listed in table 3A.1.
Note: AESZ = agro-ecological subzone; NAR = nutrient adequacy ratio. NARs for each AESZ in 2000 are calculated by dividing energetic/nutritive intake (calories and proteins) from major crop production by recommended population intake.

Table D.4 NAR per AESZ, 2020 (Regional Climate Model)

AESZ	Cassava NAR Kcal	Cassava NAR Prot.	Maize NAR Kcal	Maize NAR Prot.	Millet NAR Kcal	Millet NAR Prot.	Rice NAR Kcal	Rice NAR Prot.	Sorghum NAR Kcal	Sorghum NAR Prot.	Yam NAR Kcal	Yam NAR Prot.	All 6 crops NAR Kcal	All 6 crops NAR Prot.
1	0.00	0.00	0.05	0.09	0.22	0.37	0.03	0.04	0.27	0.47	0.04	0.03	0.61	0.99
2	0.06	0.03	0.04	0.06	0.48	0.82	0.07	0.09	0.32	0.56	0.13	0.10	1.10	1.66
3	0.85	0.42	0.25	0.39	0.00	0.00	0.04	0.05	0.16	0.28	0.32	0.24	1.61	1.38
4	0.43	0.21	0.28	0.44	0.02	0.03	0.06	0.07	0.13	0.23	0.20	0.15	1.11	1.14
5	0.06	0.03	0.09	0.14	0.06	0.11	0.00	0.00	0.20	0.34	0.05	0.04	0.46	0.66

table continues next page

Table D.4 NAR per AESZ, 2020 (Regional Climate Model) *(continued)*

	Cassava		Maize		Millet		Rice		Sorghum		Yam		All 6 crops	
AESZ	NAR Kcal	NAR Prot.	NAR Kcal	NAR Prot.	NAR Kcal	NAR Prot.	NAR Kcal	NAR Prot.	NAR Kcal	NAR Prot.	NAR Kcal	NAR Prot.	NAR Kcal	NAR Prot.
6	0.17	0.08	0.00	0.00	0.00	0.00	0.00	0.00	0.00	0.00	0.43	0.33	0.60	0.41
7	0.00	0.00	0.14	0.23	0.73	1.24	0.00	0.00	1.77	3.10	0.00	0.00	2.65	4.57
8	0.66	0.33	0.25	0.39	0.11	0.18	0.40	0.51	0.48	0.84	0.86	0.66	2.75	2.91
9	0.35	0.17	0.22	0.35	0.16	0.28	0.14	0.17	0.31	0.55	0.34	0.26	1.52	1.78
10	0.32	0.16	0.57	0.91	0.26	0.43	0.37	0.47	1.06	1.86	0.80	0.61	3.37	4.44
11	1.95	0.96	0.25	0.39	0.06	0.11	0.27	0.34	0.27	0.47	1.20	0.92	4.00	3.19
12	0.11	0.05	0.05	0.07	0.00	0.00	0.02	0.02	0.00	0.00	0.06	0.04	0.23	0.19
13	0.35	0.17	0.06	0.09	0.00	0.00	0.02	0.02	0.00	0.00	0.32	0.24	0.75	0.53
14	0.14	0.07	0.06	0.10	0.00	0.00	0.01	0.02	0.00	0.00	0.06	0.04	0.28	0.24
15	0.25	0.12	0.05	0.08	0.00	0.00	0.01	0.01	0.00	0.00	0.12	0.09	0.44	0.31
Total	0.22	0.11	0.09	0.14	0.12	0.21	0.06	0.07	0.15	0.27	0.19	0.14	0.83	0.94

NAR Kcal = Tot_production (Tons) * (nutritional value (Kcal per 100 grams) *10 * 1,000)/(population * (2,380 * 365)).
NAR proteins = Tot_production (Tons) * (nutritional value (grams per kg) *1,000) / (population * (40 * 365)).
Source: Authors' calculations based on data sources listed in table 3A.1.
Note: AESZ = agro-ecological subzone; NAR = nutrient adequacy ratio. NARs for each AESZ in 2020 are calculated by dividing energetic/nutritive intake (calories and proteins) from major crop production by recommended population intake.

Table D.5 NAR per AESZ, 2050 (Regional Climate Model)

	Cassava		Maize		Millet		Rice		Sorghum		Yam		All 6 crops	
AESZ	NAR Kcal	NAR Prot.	NAR Kcal	NAR Prot.	NAR Kcal	NAR Prot.	NAR Kcal	NAR Prot.	NAR Kcal	NAR Prot.	NAR Kcal	NAR Prot.	NAR Kcal	NAR Prot.
1	0.00	0.00	0.03	0.04	0.09	0.15	0.01	0.02	0.11	0.20	0.02	0.01	0.26	0.42
2	0.03	0.02	0.02	0.03	0.22	0.38	0.03	0.04	0.16	0.28	0.07	0.05	0.53	0.79
3	0.43	0.21	0.14	0.22	0.00	0.00	0.02	0.03	0.09	0.17	0.17	0.13	0.85	0.75
4	0.15	0.08	0.10	0.16	0.01	0.01	0.02	0.03	0.05	0.09	0.07	0.06	0.41	0.43
5	0.03	0.01	0.04	0.06	0.02	0.04	0.00	0.00	0.09	0.16	0.02	0.02	0.20	0.28
6	0.10	0.05	0.00	0.00	0.00	0.00	0.00	0.00	0.00	0.00	0.27	0.20	0.37	0.26
7	0.00	0.00	0.09	0.15	0.41	0.70	0.00	0.00	1.07	1.88	0.00	0.00	1.58	2.73
8	0.32	0.16	0.12	0.20	0.06	0.09	0.19	0.25	0.27	0.47	0.41	0.31	1.37	1.48
9	0.15	0.07	0.10	0.16	0.07	0.12	0.06	0.08	0.15	0.27	0.16	0.12	0.69	0.82
10	0.19	0.10	0.33	0.52	0.14	0.24	0.21	0.27	0.72	1.27	0.51	0.39	2.11	2.78
11	1.18	0.58	0.16	0.25	0.03	0.06	0.17	0.22	0.16	0.29	0.78	0.60	2.48	1.99
12	0.04	0.02	0.01	0.02	0.00	0.00	0.00	0.01	0.00	0.00	0.02	0.01	0.07	0.06
13	0.15	0.07	0.03	0.04	0.00	0.00	0.01	0.01	0.00	0.00	0.14	0.11	0.33	0.24
14	0.05	0.02	0.02	0.04	0.00	0.00	0.01	0.01	0.00	0.00	0.02	0.02	0.10	0.08
15	0.11	0.05	0.02	0.04	0.00	0.00	0.00	0.01	0.00	0.00	0.05	0.04	0.19	0.14
Total	0.09	0.04	0.04	0.06	0.05	0.08	0.02	0.03	0.06	0.11	0.08	0.06	0.34	0.39

NAR Kcal = Tot_production (Tons) * (nutritional value (Kcal per 100 grams) *10 * 1,000) / (population * (2,380 * 365)).
NAR proteins = Tot_production (Tons) * (nutritional value (grams per kg) *1,000)/(population * (40 * 365)).
Source: Authors' calculations based on data sources listed in table 3A.1.
Note: AESZ = agro-ecological subzone; NAR = nutrient adequacy ratio. NARs per AESZ in 2050 are calculated by dividing energetic/nutritive intake (calories and proteins) from major crop production by recommended intake of population.

Table D.6   Average Socioeconomic Status of AESZs and MAR Results

AESZ	MAR (2000)	MAR (2020)	MAR (2050)	Travel time (hrs) to cities > 250k	Travel time (hrs) to cities > 100k	Prevalence (% pop) < $2/day	Prevalence (% pop) < $1.25/ day
Dry sub-humid high plain	1.22	0.80	0.34	3.9	3.1	77.0	64.9
Dry sub-humid plain	1.94	1.38	0.66	4.4	3.6	77.5	81.2
Humid lowland and scarpland	2.08	1.50	0.80	5.3	4.3	85.5	80.3
Humid plain	1.90	1.13	0.42	6.2	5.4	79.9	78.3
Humid plateau	0.91	0.56	0.24	4.5	4.4	93.6	69.0
Perhumid high plain	0.65	0.50	0.31	9.1	8.8	93.5	63.6
Semi-arid plain	4.85	3.61	2.16	5.2	4.9	72.7	83.8
Sub-humid central Niger-Benue trough	4.07	2.83	1.42	5.8	3.9	72.4	48.8
Sub-humid high plain	2.51	1.65	0.76	5.8	4.6	80.9	54.5
Sub-humid plain	5.14	3.90	2.45	7.1	6.7	91.0	64.4
Very humid high plain	4.81	3.59	2.24	11.0	7.5	82.0	79.3
Very humid lowland	0.40	0.21	0.07	3.7	3.3	92.4	63.4
Very humid lowland and scarpland	0.98	0.64	0.28	3.1	2.6	80.7	53.9
Very humid plain	0.46	0.26	0.09	3.5	3.2	78.9	66.7
Very humid/perhumid Niger delta	0.58	0.38	0.17	8.2	8.1	52.0	38.5
Total Nigeria	1.41	0.89	0.36				

Source: Authors' calculations based on data sources listed in table 3A.1.
Note: AESZ = agro-ecological subzone; MAR = mean adequacy ratio.

Average values of poverty (defined by the prevalence of the population living on less than US$1.25 and US$2 a day) and access to market (hours of travel) for each AESZ can be used to assess whether the local population is advantaged or disadvantaged in terms of accessing resources to reduce yield gaps (increase agricultural efficiency) as needed to grant food security, especially where the NAR is far below 1.

Values of changes in the MAR for each AESZ can be associated with the average socioeconomic status of the population in the AESZ (table D.6). While a major drop in the MAR heightens threats to food security, long distances to market and a prevalence of poverty hamper efforts to improve agricultural productivity and reduce the effectiveness of food imports and food aid.

Estimates unveil the potential for food self-sustainability within each AESZ; food imports from other AESZs and abroad may mitigate food scarcity depending on the average measures of access to market and poverty in each AESZ (table D.6). Also, in terms of the variability of climate predictions, as simulated by GFDL and NCAR, climate change has a slightly lower impact on MAR in 2020 and 2050 than the impact simulated by the RCM alone (table D.6).

Note

1. http://faostat.fao.org/.

References

CIESIN (Center for International Earth Science Information Network), IFPRI, World Bank, and CIAT. 2011. Global Rural-Urban Mapping Project (GRUMPv1): Population Grids. Palisades, NY: Socioeconomic Data and Applications Center (SEDAC), Columbia University.

Hatløy, A., L. E. Torheim, and A. Oshaug. 1998. "Food Variety—A Good Indicator of Nutritional Adequacy of the Diet? A Case Study from an Urban Area in Mali, West Africa." *European Journal of Clinical Nutrition* 52: 891–98.

Nelson, A. 2008. *Travel Time to Major Cities: A Global Map of Accessibility.* Global Environment Monitoring Unit—Joint Research Centre of the European Commission, Ispra, Italy.

WHO. 1985. "Energy and Protein Requirements: Report of a Joint FAO/WHO/UNU Expert Consultation" (WHO Technical Report Series No. 724). Geneva.

You, L., S. Crespo, Z. Guo, J. Koo, W. Ojo, K. Sebastian, M. T. Tenorio, S. Wood, and U. Wood-Sichra. 2010. "Spatial Production Allocation Model (SPAM) 2000 Version 3 Release 2" (accessed August 31, 2011), http://MapSPAM.info.

Livestock Analysis

Two indicators of climate impact on livestock were considered:

* The temperature-humidity index (THI; Bohmanova, Misztal, and Cole 2007)
* The gross primary productivity (GPP) of vegetation.

In the first case, the THI combines the effect of temperature and relative humidity into a single value. It was observed (Vitali *et al.* 2009) that there are thresholds in the daily values of THI above which there is an abrupt increase in the number of animal deaths; but impacts on production and water requirements have also been observed (Howden, Hall, and Bruget 1999; Johnson 1989).

In this work, two THIs (THI1, Yousef 1985; THI2, National Research Council 1971) were calculated from Regional Climate Model (RCM) daily mean temperature and relative humidity data for the 1976–2005 baseline.

$$THI1 = T_{med} + 0.36 T_{dew} + 41.2 \qquad (E.1)$$

$$THI2 = (1.8 \cdot T_{med} + 32) - (0.55 - 0.0055 \cdot RH) \cdot (1.8 \cdot T_{med} - 26) \qquad (E.2)$$

where T_{med} is the mean daily temperature (°C), T_{dew} is the daily dew point temperature (°C), and RH the relative humidity.

THI values obtained using the same procedure for the short- and medium-term future periods were then compared with the baseline for the agro-ecological subzones (AESZs). In all cases, THI values were classified to indicate different conditions for the livestock, from no discomfort to emergency:

* THI < 68   No discomfort
* 68 ≤ THI < 72   Mild discomfort
* 72 ≤ THI < 75   Discomfort
* 75 ≤ THI < 79   Alert
* 79 ≤ THI < 84   Danger
* THI ≥ 84   Emergency

In the second case, to investigate how climate can affect GPP, and thus livestock nutrition, a statistical analysis was formulated relying on RS-based GPP product (MOD17A2[1]) using MODIS sensors, which make it possible to analyze an extended time frame (10 years).

A total of 46 (8 day syntheses per year) × 4 (tiles composing Nigeria) GPP maps from 2000 to 2010 were acquired,[2] with their quality-check images. The land cover product MOD12Q1, produced yearly by MODIS imagery and available at 500 m resolution, was also collected for available years from 2001 to 2009. This helped to restrict the analysis to grassland and savanna, the land cover categories that, e.g., provide pasturage for livestock.

The methodology consisted first in mosaicking GPP images for each date, averaging qualitatively valid pixels into seasonal maps for multiyear intervals, and then evaluating their relationships with climate conditions using a multinomial regression. Precipitation (prec) and minimum and maximum temperature (t_{min} and t_{max}) were selected as independent climate variables, from Climate Research Unit (CRU) data for 2000–06, to explain the values of GPP.

The statistical model parameters were set up at seasonal scale, then changes in GPP using RCM and perturbed outputs vs. the baseline were calculated, using these equations:

$$GPP_{djf} = 357.369 + 0.686*prec - 19.372*t_{max} + 20.136*t_{min} \qquad (R^2 = 0.76)$$
$$(E.3)$$

$$GPP_{mam} = 1310.466 + 0.031*pre - 35.930*t_{max} + 7.145*t_{min} \qquad (R^2 = 0.67)$$
$$(E.4)$$

$$GPP_{jja} = 721.968 + 0.053*prec - 0.998*t_{max} - 24.676*t_{min} \qquad (R^2 = 0.54)$$
$$(E.5)$$

$$GPP_{son} = 1036.550 + 0.151*prec - 20.999*t_{max} - 8.656*t_{min} \qquad (R^2 = 0.72)$$
$$(E.6)$$

Notes

1. https://lpdaac.usgs.gov/products/modis_products_table/mcd12q1.
2. Seven images are lacking for 2000 (January–first half of February) for two tiles over Nigeria, and one is lacking for 2001 (end of June) for three tiles.

References

Bohmanova, J., I. Misztal, and J. B. Cole. 2007. "Temperature-Humidity Indices as Indicator of Milk Production Losses due to Heat Stress." *Journal of Dairy Science* 90: 1947–56.

Howden, S. M., W. B. Hall, and D. Bruget. 1999. "Heat Stress and Beef Cattle in Australian Rangelands: Recent Trends and Climate Change." In *People and Rangelands: Building the Future*, edited by D. Eldridge and D. Freudenberger, 43–45. Proceedings of the VI International Rangeland Congress, Townsville, Queensland.

Johnson, H. D. 1989. "The Lactating Cow in the Various Ecosystems: Environmental Effects on Its Productivity." In *Feeding Dairy Cows in the Tropics*, edited by A. Speedy and R. Sansoucy. Proceedings of the FAO Expert Consultation held in Bangkok, Thailand, July 7–11. FAO Animal Production and Health Paper 86, FAO, Rome.

National Research Council. 1971. *A Guide to Environmental Research on Animals*. Washington, DC: National Academy of Sciences.

Vitali, A., M. Segnalini, L. Bertocchi, U. Bernabucci, A. Nardone, and N. Lacetera. 2009. "Seasonal Pattern of Mortality and Relationships between Mortality and Temperature Humidity Index in Dairy Cows." *Journal of Dairy Sciences* 92: 3781–90.

Yousef, M. K. 1985. *Stress Physiology in Livestock*. Boca Raton, FL: CRC Press.

Hydrological Analysis

Set-Up and Application of the ArcSWAT Model

ArcSWAT is a long-term, physically based, continuous simulation watershed model for quantifying the impact of land management practices in large catchments; it can simulate the complexity of processes of water balance. ArcSWAT makes it possible to include climatic/gauge stations across physiographic regions and multiple dams and reservoirs, and it can be calibrated and validated over large model domains.

The model divides a watershed into sub-basins, which allows for accounting of land uses and soil properties impact on hydrology. Then the model subdivides these sub-basins into smaller homogenous units, the hydrological response units (HRU), characterized by unique features of land cover and soil and its management.

The model requires specific data for simulating the water budget. The data can be categorized as spatial (Digital Elevation Model [DEM], stream network data, and land use and soil maps) and nonspatial, related to climate and discharge. Concerning climate, the model requires information on: (1) precipitation; (2) daily maximum and minimum temperatures; (3) solar radiation; (4) wind speed; and (5) relative humidity. Discharge data are necessary for sensitivity analysis, calibration, and validation.

In the spatial data, besides information on elevation and land cover, necessary soil attributes refer to structural (e.g., texture, gravel content), physical (density), biological (e.g., carbon content), and hydrologic (e.g., hydraulic conductivity, content of available water) characteristics. When they are not available, the parameters can be derived using "pedo-transfer" functions (e.g., Saxton and Rawls 2006).

Of the ArcSWAT parameters, the most important to this study, focusing on the shallow subsurface and surface water cycle, are

- GWqmn: Threshold depth of water in the shallow aquifer required for return flow to occur (mm H_2O)

- CN2: A moisture condition. An HRU with a small CN2 value will have more infiltration than one with a large value.
- ESCO: Soil evaporation compensation factor
- GW_revap: Groundwater revaporation coefficient
- ALPHA_BF: A base flow recession constant that is a direct index of groundwater flow response to changes in recharge. Values are lower in lands with slow response.
- SOL_AWC: Available water capacity of the soil layer (mm H_2O/mm soil)
- SOL_Z: Depth from soil surface to bottom of layer (mm).

After consultation with local experts, input data were selected to support the hydrological analysis.

Spatial Inputs

Of the spatial data the most important is the DEM, which is useful for breaking the territory into physiographic units. Because it is pan-national and considering the resolution (about 8 km) of climate input to be used in the final analysis, this study used the latest version of the DEM from the SRTM,[1] and resampled to 1 km resolution.

Given the necessity of hydrologically coherent correspondence between elevation, stream network, and watershed boundaries to be modeled, using the digitized stream network or even what is automatically extracted using other elevation datasets is not possible. So the DEM chosen has been used to partition the Nigerian territory into hydrologically connected sub-basins. The most reliable terrain analysis procedures have been used to overcome the known limit of the standard procedures used incorporated into many Geographic Information System (GIS)-based hydrological tools (Nardi *et al.* 2008). This means that the procedures of Tarboton (1997) and Tarboton, Bras, and Rodriguez-Iturbe (1991), rather than the one embedded into ArcSWAT, were selected for reproducing the flow directions, drained areas, and extract stream network to feed the model.

With these procedures 893 sub-basins completely falling inside Nigerian boundaries were extracted, jointly with their reach (map F.1). It was decided to rely on physiographic units even if covering the whole national territory was not covered because the national administrative units for planning and the natural watershed units do not match. Because policy is an outcome of political, not hydrological, processes, solutions for water problems in trans-boundary basins, most of which fall outside Nigeria, are mainly influenced by decisions made by other countries.

For each watershed delineated, topographic statistics useful to ArcSWAT have been calculated (e.g., mean slope, stream length); in particular the slope was classified into five categories (table F.1). The classes were extracted from the slope frequency distribution using a Jenks natural break algorithm to reduce variance within classes and maximize variance between classes.

A second spatial dataset is related to land cover to take into account the national scale of the analysis, the need to incorporate land cover classifications in

Map F.1 Hydrological Areas of Nigeria by Rivers (Left) and Sub-basins (Right)

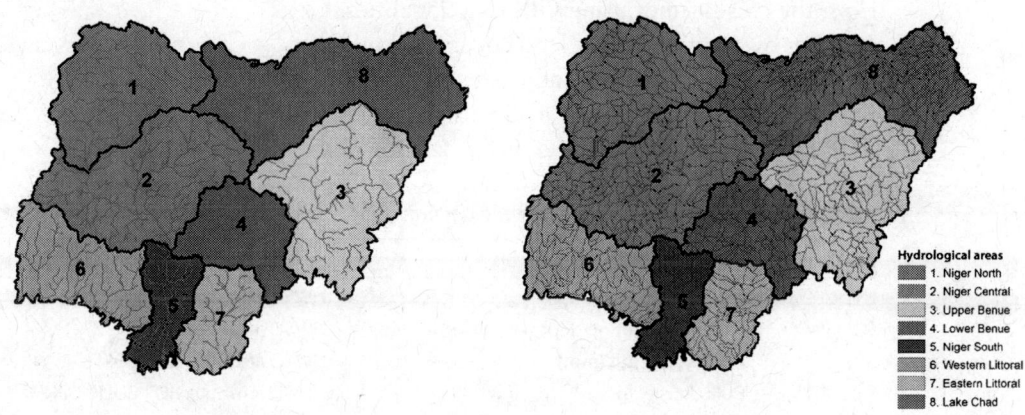

Hydrological areas
1. Niger North
2. Niger Central
3. Upper Benue
4. Lower Benue
5. Niger South
6. Western Littoral
7. Eastern Littoral
8. Lake Chad

Source: ArcSWAT.

Table F.1 Value Intervals Used to Extract Slope Classes

Slope class	Slope range (%)
1	< 1.14
2	1.14–4.18
3	4.18–9.89
4	9.89–18.64
5	> 18.64

ArcSWAT, and the necessity to have pixels with a single type of land cover. Since 2005 is the reference year for the end of historical analyses and the beginning of simulations of the future, the MOD12Q1[2] land cover map produced from MODIS images for 2006 (MOD2006) and the Global Land Cover dataset 2005–06 (GLC2006) were merged to fill reciprocal gaps and then feed ArcSWAT. To combine the two datasets, they were first resampled from their original resolution of 500 m (MOD2006) and 300 m (GLC2006) to 1 km. In both cases the "majority" algorithm was chosen in order to associate to each coarser resolution (1 km) pixel the most frequent land cover among those of the pixels at finer resolution included. Then, a SWAT code was associated to each IGBP[3] land cover class of MOD2006 grid.

Class 14 for IGBP classification (cropland/natural vegetation mosaic) is not considered in ArcSWAT classification system; this class was thus better detailed using the GLC2006 classification system.

This cross-merging also made it possible to fill voids at the remaining NOCL (not classified) small clusters of pixels like water and bare lands.

Concerning soil, the best available dataset is the HWSD[4] at 1 km resolution. According to ArcSWAT documentation, the attributes reported in table F.2 were associated with each of 174 soil units falling in Nigeria, in some cases through a simple reclassification, in other cases using pedotransfer functions (Saxton and

Rawls 2006). For non-soil units (lakes) the lake-bottom soil was reconstructed from the closest ones using GIS functionalities.

After overlapping slope, land cover, and soil layers, each sub-basin was divided into 15,338 distinctive combinations of land units, the hydrological response units (HRUs; map F.2) and assumed having specific characteristics in terms of their behavior in affecting the water balance processes.

Table F.2 Attributes Parameterized for Each Soil Unit

SWAT code	Description	Sources
HYDGRP	Soil hydrologic group	HWSD attribute: DRAINAGE
SOL_ZMX	Maximum rooting depth of soil profile (mm)	HWSD attribute: ROOTS
SOL_Z	Soil layer depth (mm)	HWSD attribute: REF_DEPTH
SOL_BD	Bulk density (g/cm^3)	HWSD attribute: REF_BULK_DENSITY
SOL_AWC	Available water content (mm H2O/mm soil)	HWSD attribute: AWC
SOL_K	Saturated hydraulic conductivity (mm/hr)	Saxton and Rawls (2006) from HWSD
SOL_CBN	Organic carbon content (%)	HWSD attribute: OC
CLAY	Clay content (%)	HWSD attribute: CLAY
SILT	Silt content (%)	HWSD attribute: SILT
SAND	Sand content (%)	HWSD attribute: SAND
ROCK	Rock fragment content	HWSD attribute: GRAVEL
SOL_ALB	Soil albedo	Ten Berge (1986) from HWSD
USLE_K	USLE soil erodibility	SWAT documentation using HWSD inputs

Map F.2 Slope Classes (a), Land Cover Categories (b), and Soil Types (c) Used to Extract Hydrological Response Units (HRUs) (d)

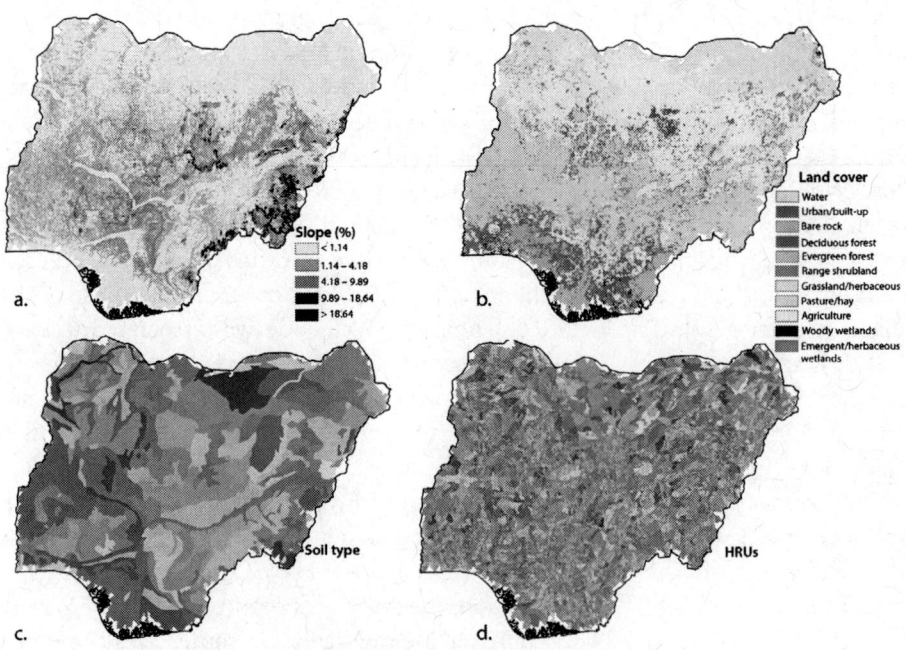

Source: Authors' calculations based on data sources listed in table 3A.1.

Nonspatial Inputs

Climate data are necessary for running the water balance computation in ArcSWAT. For this purpose, according to the reliability of station position after cross-checking their coordinates between Nigerian Meteorological Agency (NIMET) and National Climate Data Center (NCDC) datasets, 29 stations were selected for rainfall and 21 for minimum/maximum temperature, distributed as in map F.2 and combined with hydrological area (HA) and sub-basin maps. Daily rainfall and minimum/maximum temperature series were used directly to feed ArcSWAT. Given the large gaps in such records, ArcSWAT has a weather generator for reconstructing meteorological series, starting from long-term monthly statistics (most on precipitation but also on minimum and maximum temperatures, solar radiation, number of rainy days, wind speed, and dew point temperature as a proxy for relative humidity). The weather generator was set up using station data for precipitation and temperature; for other variables the Regional Climate Model (RCM)-generated variables were used.

Besides supplying precipitation inputs and other meteorological variables for computing the evaporation and evapotranspiration components, rainfall data are useful for calibrating the model; indeed, the simulated streamflow series (the water flow for superficial runoff and groundwater return) can be compared with observed records at suitable locations.

Discharge information was extensively researched by analyzing datasets from the Integrated Water Resources Project (only for HA8); *the Hydrological Year Books* from Nigeria Hydrological Services Agency (NIHSA); and the Japan International Cooperation Agency (JICA) National Water Resources Master Plan (NWRMP) (1995). The last one is the most complete in terms of HAs and time coverage (89 stations for different time periods between 1960 and 1989), and it was the basis for selecting stations to support sensitivity analysis and calibration of the model (see below); the discharge station locations are shown in map F.3.

In making simulations, ArcSWAT associates to each HRU the closest meteorological station from which data are collected. This is a serious limitation because the station often represents an area covering tens of kilometers and the elevation is not necessarily representative, which is crucial for interpolating meteorological data. This and other limits due to the reciprocal position of rainfall and discharge stations will be better discussed below with calibration.

ArcSWAT: Sensitivity Analysis, Calibration, and Validation

In the initial phase, a calibration procedure was performed for six HAs to seek a compromise between the availability of precipitation vs. discharge data from measurement stations. Indeed many difficulties arose because different time frames were covered by precipitation (1975–2009) and stream flow (1960–89) station data, both had large inter- and intra-annual gaps, some locations were uncertain, and information about data reliability was lacking.

Temporal gaps in meteorological station data were addressed by using a weather generator; this made it less reliable to use, for example, "abnormal"

Map F.3 Measurement Stations for Rainfall-Temperature and Discharge (JICA Master Plan) in the Hydrographic Network

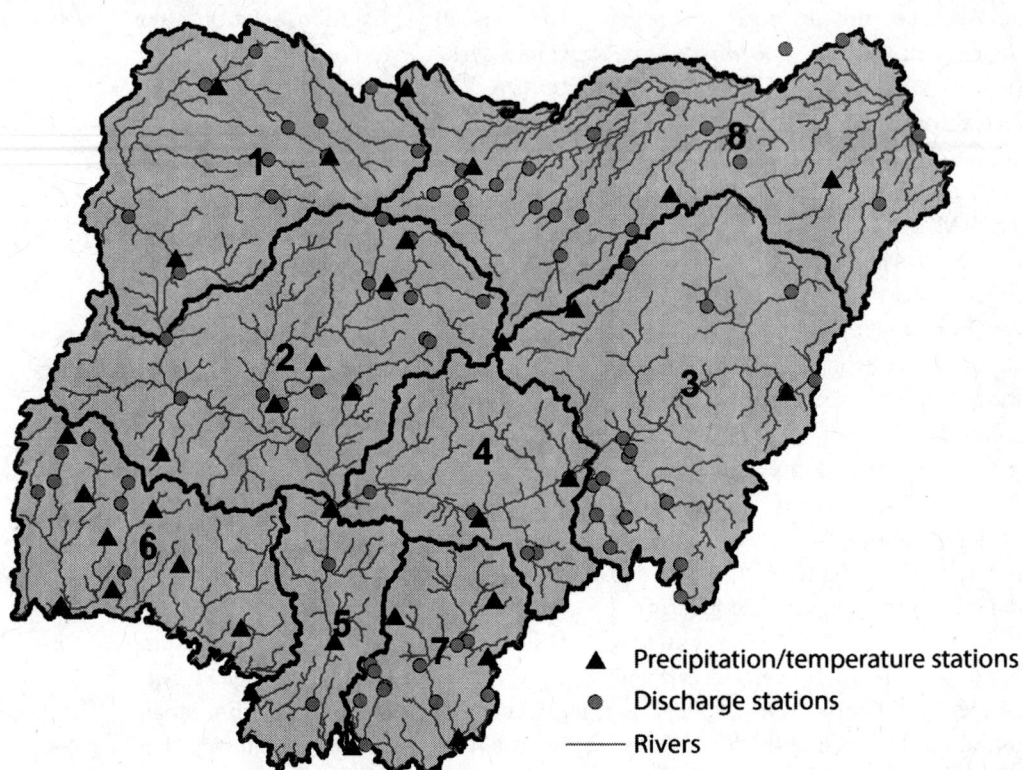

▲ Precipitation/temperature stations
● Discharge stations
─── Rivers

Sources: NIMET for rainfall-temperature information; JICA1995 for discharge information.
Note: Numbers refer to hydrological areas. JICA = Japan International Cooperation Agency.

discharge years (when there was unusual deviation from monthly series) for calibration when the gaps were present in precipitation time series. Indeed in case of gaps the weather generator reproduces the climate according to 30-year statistics, and may not be able to consistently support rainfall-runoff simulation for extraordinary years.

In addition, lack of certainty about the location of both types of stations causes errors of several kilometers. Further, meteorological stations available for calibration (those closest to the basin), often are located far from the streamflow stations and even downstream, so they would represent the rainfall-runoff spatio-temporal dynamics well.

As calibration was possible for selected basins inside the HAs according only to streamflow station best-estimated positions, this also limited parameters to ones strictly related to the runoff and superficial-shallow soil water dynamics (e.g., neither management nor deep groundwater parameters could be considered because no detailed spatialized data were available). Other soil parameters were considered reliable (since the most updated and spatially resolved dataset, the Harmonized World Soil Dataset, was used) and adjusted only if considered significant after the model sensitivity analysis described next.

Despite the limitations, the procedure was as much as possible optimized, and cross-checking led to the selection of at least one station for each HA to be used in calibration. Table F.3 lists stations for which calibration was initially tested for several HAs (see also map F.4).

All streamflow data for stations functioning for only a few years, having diffuse intra- and inter-year gaps, or not located along river networks or at dams, were excluded. On choosing the temporal frame to be used for calibration/validation, it was also considered intuitively that the most recent observations are based on more reliable technology instrumentation for both variables, so the 1980s were preferred when accessible.

Of the more than 60 parameters in the SWAT model, some vary by sub-basin, land use, or soil type, which increases the true number of parameters substantially. In any case, the first step of calibration was, for each HA, a sensitivity analysis to select parameters to be calibrated.

Given the focus on water availability and the wide spatial extent of the analysis, and considering the work of van Griensven *et al.* (2006), the analysis ranked the most sensitive parameters. Of these, groundwater runoff was ranked highly important, and the next five—evapo(transpi)ration, soil and sub-soil (shallow groundwater) processes—important, and thus influence the hydrology of the system which is the focus of our analysis. Table F.4 shows how, although with slightly different ranking, the most sensitive parameters are confirmed across HAs, which also validates the results of Schuol *et al.* (2008).

Because the Banaga station is at the junction of two tributaries, its discharge series were used in calibrating two adjacent sub-basins.

The calibration exploited the ability of the SWAT-CUP[5] software package, in particular of the SUFI-2 module, to perform calibration/validation and offer a large choice of objective functions. In addition, in a comparison study

Table F.3 Discharge Stations Initially Evaluated for Calibration Purposes

HA	Discharge station	Continuous years of discharge observation	Closest meteorological station
1	**Banaga**	**1970–84**	**Gusau**
2	Kaduna south	1973–77 and 1979	Kaduna
	Zaria-Kano	1973–76 and 1979–80	Zaria
	Malendo	1973–78	Kaduna
	Komi	**1984–89**	**Yelwa**
3	**Dadinkowa**	**1970–75 and 1982–89**	**Bauchi**
	Gindin Dorowa	1981–85	Ibi
	Donga	1970 and 1974–76	Ibi
4	**Katsina Ala**	**1970–72 and 1974–84**	**Makurdi**
8	**Gwarzo**	**1965–67, 1969–79, 1981–82, and 1984–89**	**Kano**
	Challawa	1964–89	Kano
	Gashua	1964–89	Nguru
	Tiga	1964–85	Kano
	Birnin Kudu	1964–67, 1971–72, 1997, and 1981–83	Kano

Note: HA = hydrological area. Boldfaced stations produced the most satisfying results and were used for this study and are shown in map F.4.

Map F.4 Stations with Streamflow Observations

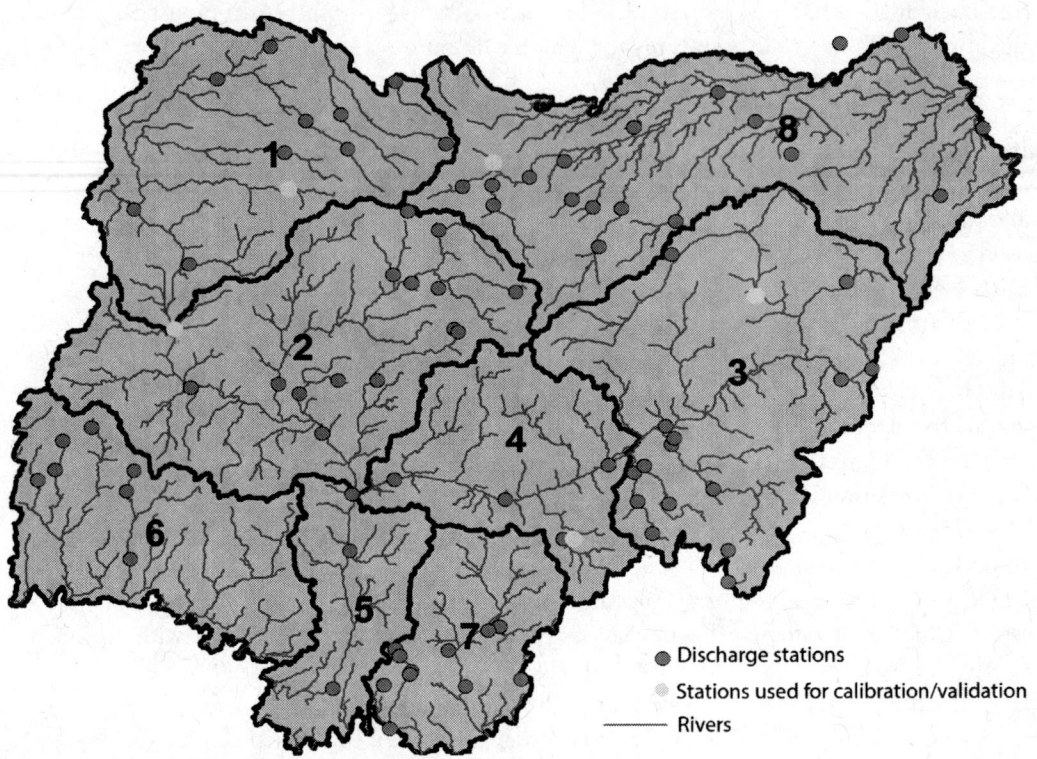

Source: Authors' calculations based on data sources listed in table 3A.1.
Note: Numbers refer to hydrological areas.

Table F.4 HAs and the Most Important Parameters for SWAT Calibration[a]

HA	GWqmn	CN2	ESCO	SOL_AWC	SOL_Z	GW_revap	ALPHA_BF
1	1	2	3	4	5	6	—
2	1	4	2	3	5	6	—
3	1	2	3	4	5	—	6
4	1	3	2	5	6	—	4
7	1	5	2	4	6	—	3
8	1	3	2	4	5	6	—

Note: HA = hydrological area; SWAT = Soil and Water Assessment Tool. Acronyms in the column headings are explained on pages 145–146. — = not available.
a. Numbers in columns refer to importance, with 1 the most important.

Yang *et al.* (2007) found that SUFI-2 required much fewer simulations than other methods while producing similar Nash-Sutcliff and R^2 values when the best calibration and validation results were compared with measured data.

For each of the HA sub-basins selected, multiple calibration runs were performed, combining two different objective functions NS (Nash and Sutcliffe 1970) for 50 iterations and coefficient of determination R^2 multiplied by the coefficient b of the regression line (BR^2), with 100 iterations. Considering the constraints already discussed, and following conventions in hydrologic modeling, discharge data for at least two continuous "normal" years were identified and data for the remaining year were used for validation.

Table F.5 reports the new range of parameters derived from these multiple runs. The one from the best-performing run in terms of *P-factor* (the percentage of measured data bracketed by the 95 percent prediction uncertainty; 95PPU) and *R-factor* (the average thickness of the 95PPU band divided by the standard deviation of the measured data) was kept for successive simulations. SUFI-2 thus seeks to bracket most of the data measured (large *P-factor*, maximum 100 percent) with the smallest possible value of *R-factor* (minimum 0, with values around 1 being reasonable).

In discarding iterations as not useful, it was also considered where streamflow peaks were not well reproduced in terms of timing rather than volume. Figure F.1 shows monthly series for periods, including years considered, for validation for selected HAs.

Table F.5 New Ranges and Values of Top Seven SWAT Parameters after Calibration

HA	GWqmn	CN2	ESCO	SOL_AWC	SOL_Z	GW_revap	ALPHA_BF
1	−69.47:1177.29	−9.94:12.93	0.24:0.65	−11.18:16.68	−10.03:50.12	0:0.18	—
	(660)	(2.25)	(0.395)	(2.75)	(23.45)	(0.7)	
2	−236.59:1316.59	−37.43:4.43	0.40:1.22	−1.91:44.91	−11.41:16.41	0.04:0.25	—
	(540)	(−16.5)	(0.81)	(21.5)	(2.5)	(0.17)	
3	−97.24:1737.24	−29.92:6.92	−97.24:1737.24	−1.42:46.42	−1.15:34.72	—	0.5:15
	(820)	(−11.50)	(0.57)	(22.50)	(18.32)		(7.2)
4	−1317.93:237.93	−46.41:1.41	0.42:1.28	−2.40:26.42	−3.14:14.37	—	0.32:0.98
	(540)	(−22.5)	(0.85)	(12.80)	(9.14)		(0.65)
8	−238.58:1318.58	−17.93:10.93	−0.28:0.58	0.4:15.3	0.05:78.3	0:0.31	—
	(540)	(−3.5)	(0.15)	(11.28)	(26.54)	(0.25)	

Note: — = not available; SWAT = Soil and Water Assessment Tool. Acronyms in the column headings are explained on pages 145–146.

Figure F.1 Simulated and Observed Monthly Streamflow Values for Katsina Ala Station (HA4) for 1980 and 1983–84 (Validation Period) and 1981–82 (Calibration Period)

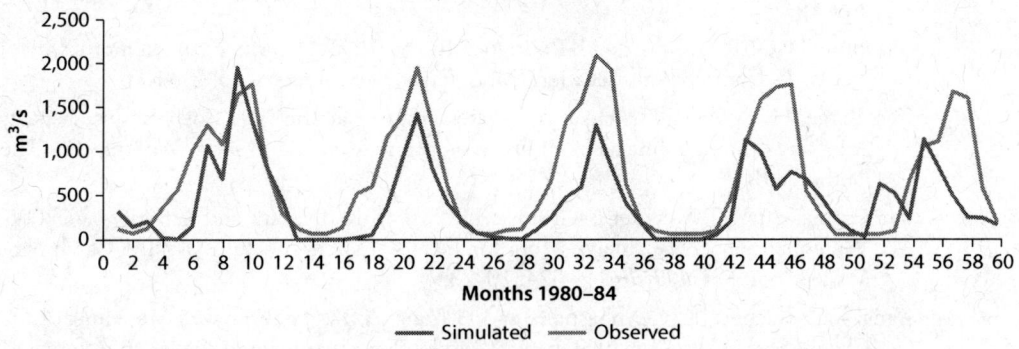

Source: Authors' calculations based on data sources listed in table 3A.1.

In the second phase of the project, since it was impossible to calibrate the remaining areas, their similarity with other HAs in terms of HRU (composition of land cover, slope, and soil type) was evaluated. When there was hesitancy about two or more similar HAs, the goodness (reliability) of completed calibrations was also considered. Using this approach, HA6 was assimilated to HA2, and HAs 5 and 7 to HA4.

Notes

1. Shuttle Radar Topographic Mission; http://srtm.csi.cgiar.org/.
2. https://lpdaac.usgs.gov/products/modis_products_table/mcd12q1.
3. http://edc2.usgs.gov/glcc/globdoc2_0.php#app2.
4. Harmonized World Soil Dataset; http://www.iiasa.ac.at/Research/LUC/External-World-soil-database/HTML/.
5. http://www.eawag.ch/forschung/siam/software/swat/index_EN.

References

JICA (Japan International Cooperation Agency). 1995. *The Study on the National Water Master Plan*. Sector Report Vol. 2. Report prepared for the Federal Ministry of Water Resources and Rural Development, Abuja.

Nardi, F., S. Grimaldi, M. Santini, A. Petroselli, and L. Ubertini. 2008. "Hydrogeomorphic Properties of Simulated Drainage Patterns Using DEMs: The Flat Area Issue." *Hydrological Sciences Journal* 53 (6): 1176–93.

Nash, J. E., and J. V. Sutcliffe. 1970. "River Flow Forecasting through Conceptual Models 1. A Discussion of Principles." *Journal of Hydrology* 10 (3): 282–90.

Saxton, K. E., and W. J. Rawls. 2006. "Soil Water Characteristic Estimates by Texture and Organic Matter for Hydrologic Solutions." *Soil Science Society of America Journal* 70 (5): 1569–78.

Schuol, J., K. C. Abbaspour, R. Sarinivasan, and H. Yang. 2008. "Estimation of Freshwater Availability in the West African Sub-Continent Using the SWAT Hydrologic Model." *Journal of Hydrology* 352 (1–2): 30–49.

Tarboton, D. G. 1997. "A New Method for the Determination of Flow Directions and Upslope Areas in Grid Digital Elevation Models." *Water Resources Research* 33: 309–19.

Tarboton, D. G., R. A. Bras, and I. Rodriguez-Iturbe. 1991. "On the Extraction of Channel Networks from Digital Elevation Data." *Hydrological Processes* 5: 81–100.

Ten Berge, H. F. M. 1986. "Heat and Water Transfer at the Bare Soil Surface: Aspects Affecting Thermal Imagery." PhD thesis, Agricultural University Wageningen, The Netherlands.

van Griensven, A., T. Meixner, S. Grunwald, T. Bishop, M. Diluzio, and R. Srinivasan. 2006. "A Global Sensitivity Analysis Tool for the Parameters of Multi-variable Catchment Models." *Journal of Hydrology* 324: 10–23.

Yang, J., P. Reichert, K. C. Abbaspour, and H. Yang. 2007. "Hydrological Modelling of the Chaohe Basin in China: Statistical Model Formulation and Bayesian Inference." *Journal of Hydrology* 340: 167–82.

Case Study Sites for the Hydrological Analysis

Table G.1 Characteristics of Hydropower, Irrigation, and Multipurpose Schemes Selected for the Study

Site number	Project	Hydrological area	Purpose and classification	Features
1	Shiroro	Niger Central	Hydropower (600 MW). Large power scheme	The Shiroro hydroplant is located on River Kaduna in north central Nigeria. The reservoir capacity is 7,000 mm³. The plant has a capacity of 600 MW and a firm power of 1,660 GWh annually.
2	Zungeru	Niger Central	Hydropower (400 MW). Large power scheme	Zungeru Dam is located on the Kaduna River downstream of the Shiroro Dam. Zungeru Dam is proposed to be optimized at 88 m, with dam height corresponding to elevation +230 m at full supply level. Firm power yield is 2,130 GWh annually.
3	Gurara	Niger Central	Water supply Irrigation Hydropower (30 MW). Medium power scheme	The Gurara multipurpose dam is near Jere in Kaduna State on the Gurara River. Reservoir capacity is 880 mm³ and is meant to transfer water (at a rate of 8 m³/s) through a 75 km tunnel to the Lower Usuma Dam in FCT for water supply; generate 30 MW in hydropower; and irrigate downstream areas using the penstock discharge (28 m³/s). The current area irrigated is about 2,000 ha. Water transfer at a rate of 8 m³/s to FCT is a priority, and hydropower operation depends on the reservoir storage.
4	Mambilla (Gembu)	Upper Benue	Hydropower (2,600 MW). Large power scheme	The Mambilla hydropower site in the upper Donga River consists of a main regulating reservoir with a dam at Gembu and smaller reservoirs at Sumsum and Nghu for daily regulation. At full supply level of +1,296 m (82 m dam height), the installed capacity is 2,600 MW and firm power is 4,417 GWh annually. The head is defined by the lower Nghu dam, not the Gembu River; thus power is produced at a constant net head of 939 m, and the Gembu Reservoir contributes a safe yield of 55.9 m³/s (equivalent to 5.90829 GWh per 1 m³/s).

table continues next page

Table G.1 Characteristics of Hydropower, Irrigation, and Multipurpose Schemes Selected for the Study *(continued)*

Site number	Project	Hydrological area	Purpose and classification	Features
5	Dadinkowa	Upper Benue	Irrigation Hydropower (34 MW). Medium power scheme	The Dadinkowa dam is about 45km east of Gombe on the Gongola River. The dam has a catchment area of 36,000 km^2 and a reservoir capacity of 2,800 mm^3 with active storage of 1,770 mm^3. The dam was designed as a multipurpose project to provide irrigation water for 38,000 ha, domestic water supply of 86,400 m^3/d, and 34 MW of hydropower. The irrigation area is not developed, and the power plant is not installed. Annual firm power yield is 186 GWh.
6	Ikere Gorge	Western Littoral	Irrigation Water supply Hydropower (6 MW). Small power scheme	The Ikere Gorge Dam is a major earth-fill dam on the Ogun River in the southwest of Nigeria. Reservoir capacity is 690 mm^3, and the dam was planned to generate 6 MW of hydropower, supply water to local communities and to Lagos, and irrigate 12,000 ha of land through a sprinkler system. Construction, which began in 1982, is not yet completed. Firm power yield is 2.01 GWh/m (24.16 GWh a year), supplying 158 mm^3.
7	Tiga	Lake Chad	Irrigation Water supply Ecology requirements	The Tiga dam is on the Kano River in northern Nigeria. The water is impounded (reservoir capacity is 1,345 mm^3) for the Kano River Irrigation Scheme, water supply to Kano City and downstream through the Hadejia River, and then the Yobe River. Area covered by the Kano River Irrigation Scheme is 20,000 ha with a potential of 40,000 ha. The current population of Kano City is 6,000,000, and 50% of domestic demand is expected to be supplied from the Tiga Reservoir. Downstream demand includes ecology and irrigation requirements in the Hadejia Nguru Wetlands. Rice is the main crop irrigated; efficiency of the scheme is about 50%. There is a plan to develop the floodplain to irrigate 20,000 ha. This will increase the downstream release from 156 mm^3/yr to 252 mm^3/yr.

Note: GWh = gigawatt hour; ha = hectare; MW = megawatt.

Power Generation Model

Simulation models usually require explicit statement of operating rules, but optimization models do not. The model suggests or prescribes operations. The objectives for reservoir operations must be explicitly stated in the form of penalty functions. For reservoir systems, optimization models require mathematical constraints to represent physical, engineering, or legal constraints and representation of hydrologic inputs to the system.

The representation of hydrologic uncertainty in optimization models has been developed from several different perspectives. The two main schools of thought (USACE 1996) on this subject are explicit and implicit stochastic representation. Explicit stochastic representation of hydrologic uncertainty and variability requires characterizing hydrologic inputs in explicit probabilistic terms, i.e., joint probability functions and time series correlations. The concept of probabilistic distribution of unregulated inflows in ESO (Explicit Stochastic Optimization) makes the SDP (Stochastic Dynamic Programming) model suitable for addressing changing hydrology in reservoirs. This was adopted for the Nigeria study.

The objective function of a reservoir system (figure H.1) is expressed as:

$$f_t = B_t + B_{t-1} + \ldots \ldots B_T + f_{T+1} \tag{H.1}$$

where B_t is the return at stage t due to the release R given the initial and final storages, and f_{T+1} describes the value of water at the end of stage T, the last stage in the planning period (planning period is 12 months). The benefit is to maximize the area under irrigation, the energy generated, or the amount of water for domestic use. For single-purpose schemes, the objective criteria are to (1) maximize the energy production at energy price of ₦6 per kilowatt hour (kWh) for firm power, a penalty of ₦6 per kWh for deficit power, and a secondary price of ₦2 per kWh for secondary power; and (2) minimize water spill. For multipurpose schemes (Gurara, Dadinkowa, and Ikere Gorge), the power model was run to optimize energy production and minimize spill because water supply and irrigation needs are met.

Figure H.1 Schematic Diagram of a System

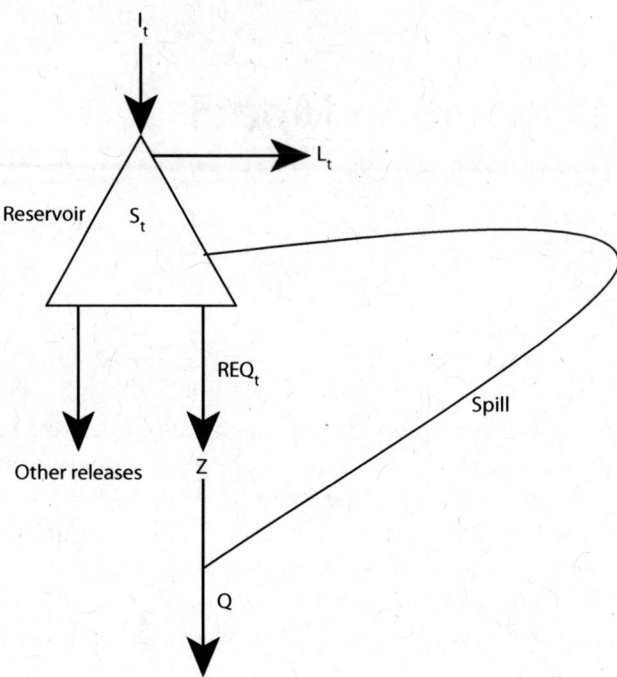

Note: I is inflow, Q is release, S is storage, L denotes losses including evaporation and seepage, REQ is the discharge through the turbine, Z is the generator, and t denotes time.

The analysis starts at time T and moves backward using Bellman's principle, which states that an optimal policy has the property that whatever the initial state and initial decisions are, the remaining decisions must constitute optimal policy for the state resulting from the first decision. The operation is subject to the following constraints:

1. Continuity equation:

$$S_{t-1} + I_t - L_t - Q_t = S_t \tag{H.2}$$

where Q_t is outflow from the reservoir, L_t is loss from the reservoir, S_t is storage, and I_t is inflow to the reservoir.
2. Storage constraint:

$$Smin < S_t < Smax_t \tag{H.3}$$

$$S_{t+1} \leq Smax_t \tag{H.4}$$

where Smin is reservoir dead capacity and $Smax_t$ is maximum storage at time t.
3. Release constraint:

$$Q_t \geq maximum(MQ_t) \tag{H.5}$$

where MQ_t is the obligatory water requirement at time t—the release from the reservoir to meet the minimum downstream demand (ecology, irrigation, and water supply).

4. Energy production, expressed as the energy production capacity (EPC):

$$EPC_t = C * REQ_t * H_t * \eta \qquad (H.6)$$

where C is the conversion potential factor for electrical energy, H is the average head over turbine, and η is energy plant efficiency. The energy that can be produced is restricted by plant capacity (PCAP) and number of hours available for energy production (NHP). Thus, the maximum peak power produced (MPEP) is

$$MPEP_t = PCAP_t * \eta * NHP_t. \qquad (H.7)$$

The power produced at any time t is

$$PKE_t = minimum(TEP_t, MPEP_t) \qquad (H.8)$$

where PKE is the peak power produced and TEP is the total power that can be produced at a particular time.

The power model (based on SDP) was used to obtain a monthly release policy for each reservoir where the state variable is reservoir storage, St, at the beginning of a stage, and the decision variable is reservoir release, Rt. The solution to the recursive equation (H.1) was obtained by working backwards in time from the end of the decision horizon (12 months). The operation model on a monthly time scale has a number of storage discretizations (not less than 15) for each reservoir. Monthly power produced was calculated using equations H.6, H.7, and H.8. The release policy and the energy generated were used to assess hydropower system reliability: Reliability or frequency of success is the ratio of the number of times the monthly power target is met to the number of months of operation.

Reference

USACE (US Army Corps of Engineers). 1996. *Developing Seasonal and Long-Term Reservoir System Operation Plans Using HEC-PRM*. Hydrologic Engineering Center, Davis, CA. www.hec.usace.army.mil.

APPENDIX I

Macroeconomic Analysis

The Intertemporal Computable Equilibrium System (ICES) model employed for the macroeconomic analysis solves recursively a sequence of static equilibria linked by endogenous investment determining the growth of capital stock from 2004 to 2050. The calibration year is 2004. This model is based on the GTAP 7 database (Narayanan and Walmsley 2008), which has been enriched to better serve the purposes of the Nigeria study. Like all other computable general equilibrium (CGE) models, ICES makes use of the Walrasian perfect competition paradigm to simulate adjustment processes, although it is also possible to include elements of imperfect competition.

Industries are modeled through a representative firm, minimizing costs while taking prices as given. Output prices are given by average production costs. The production functions are specified by a series of nested CES functions. Domestic and foreign inputs are not perfect substitutes, according to the Armington (1969) assumption.

A representative consumer in each region receives income, defined as the service value of national primary factors (natural resources, land, labor, capital). Capital and labor are perfectly mobile domestically but immobile internationally. Land and natural resources, on the other hand, are industry-specific.

The income is used to finance three classes of expenditure: aggregate household consumption, public consumption, and savings. Expenditure shares are generally fixed, which basically says that the top-level utility function has a Cobb-Douglas specification.

Public consumption is split into a series of alternative consumption items, again according to a Cobb-Douglas specification. However, almost all expenditure is actually concentrated in one industry: nonmarket services.

Similarly, private consumption is split into a series of alternative composite Armington aggregates. However, the functional specification used here is the constant difference in elasticities form: a non-homothetic function that is used to account for possible differences in income elasticities for the various consumption goods.

Investment is internationally mobile: savings from all regions are pooled and investment is then allocated so as to achieve equality of expected rates of return to capital. In this way, savings and investments are equalized at the world, but not at the regional, level. Because of accounting identities, any financial imbalance mirrors a trade deficit or surplus in each region.

The AEZ Approach

A first improvement to the original model structure was to adopt the agro-ecological zone (AEZ) approach. In its original specification, ICES assumes that within countries or regions land is allocated to different agricultural uses (crops) in response to changes in the relative prices of agricultural commodities. What governs the land-switching mechanism is an elasticity of transformation parameter that summarizes all possible economic, environmental, geographical, and biological factors constraining land uses. (This is, of course, a very rough representation of land allocation mechanisms.) The AEZ approach is a first step toward better description of land use patterns. Following Food and Agriculture Organization of the United Nations (FAO) and International Institute for Applied Systems Analysis (IIASA) methodology, world land endowment is split into 18 different AEZs (see map I.1).

The AEZ database (Avetisyan, Baldos, and Hertel 2011) identifies crop, forest extent, and production for each region by AEZs. The original data consist of detailed information for 175 crops that were aggregated into the GTAP 8 crops definition.

As the AEZs approach is embedded in ICES, land is now assumed to be suitable to different uses within, but not between, AEZs. This implies, for instance, that in a given country only crops already being cultivated in each AEZ can be

Map I.1 AEZs in the GTAP/ICES Database

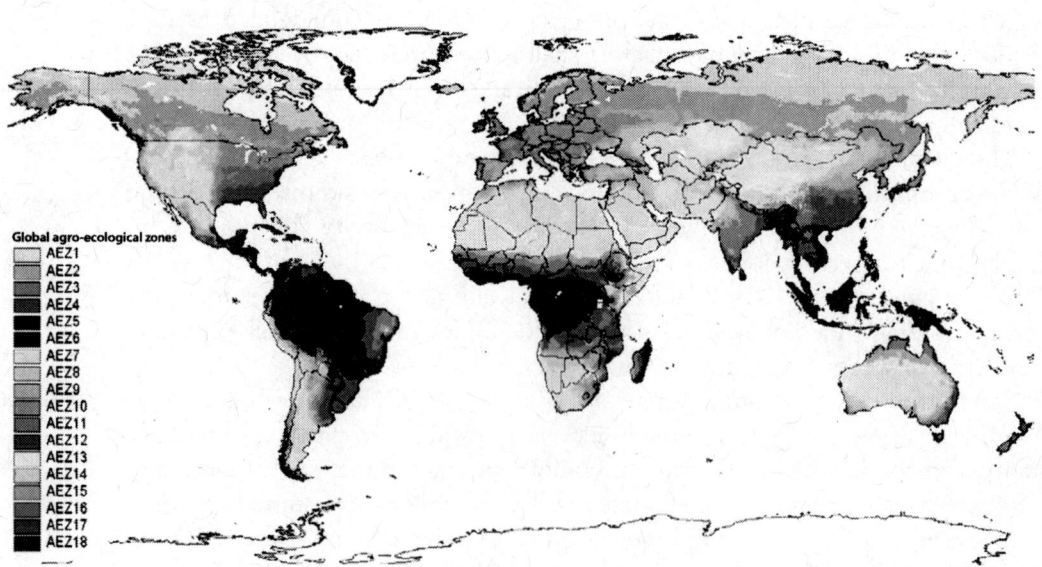

Source: Modified from Lee *et al.* 2005.

switched, and that in principle land elasticity of transformation could differ between AEZs. Therefore land substitution mechanisms are much more realistic than before, better capturing biological and geographical differences of different types of land. Nigeria is divided into six AEZs.

ICES Sectoral and Regional Disaggregation

The original GTAP 7 database incorporates 10 "agricultural industries." Eight are crop aggregates: rice, wheat, other cereals, sugar cane and beet, vegetables and fruits, plant-based fibers, and other crops.[1] Given their relevance in the Nigerian agricultural production, the detail of crops produced by the country has been increased, singling out cassava and yam. These crops are the top two in terms of share of agricultural value added, building up in 2006 to 16.3 percent for cassava and 14.7 percent for yams (Nwafor, Diao, and Alpuerto 2010), shares that are very similar to those reported in the original 2004 GTAP AEZ database (see table I.1).

The value of cassava and yam production has been disentangled from the GTAP aggregate "vegetable and fruits," to which both belong. Data for the process are drawn from compounding information from the GTAP AEZ database (Avetisyan, Baldos, and Hertel 2011), which reports the quantity produced of each crop per AEZ with the values provided by Nwafor, Diao, and Alpuerto (2010).

Because they are less relevant, other kinds of agricultural productions have been aggregated in larger bundles, to comply with the information provided for analysis of changes in crop productivity.

Figure I.1 shows the final sectoral specification of the model used in this exercise. It also reports the macrosectoral aggregations to which each sector belongs.

Because ICES is a global model, the rest of the world had to be considered, even though the assessment is concerned with Nigeria. The regional detail chosen for the model is reported by figure I.1. The choice was motivated by the need both to simplify and to keep a reasonable balance between the size of region to avoid "strange" effects from and on international trade patterns.

The Baseline

Having tailored the database for 2004, it was then a necessary step to construct a future baseline that can capture potential economic development in Nigeria up to 2050. This baseline is the counterfactual "without climate change" on top of which the impacts of climate change on crop productivity are imposed and against which the consequent gross domestic product (GDP) and sectoral performance of Nigeria's economic system will be contrasted.

Table I.1 Agricultural Value-Added of Cassava and Yams

| | Shares of production over agriculture | |
	Nwafor, Diao, and Alpuerto (2010)	GTAP 7
Cassava	16.32	17.46
Yams	14.73	15.77

Toward Climate-Resilient Development in Nigeria • http://dx.doi.org/10.1596/978-0-8213-9923-1

Figure I.1 ICES Sectoral and Regional Aggregation

Rice	
Cassava	
Yam	
Other cereal crops	**Agriculture**
Vegetables and fruits	
Other crops	
Livestock and fishing	
Timber	
Coal	
Oil	**Mining**
Gas	
Mining	
Oil products	
Electricity	**Manufacturing**
Other industries	
Private services	**Services**
Public services	

ICES regions

USA	USA
EUROPE	Europe
FSU	Former Soviet Union
RoA1	Rest of Annex 1
MENA	Middle East and North Africa
NIGERIA	Nigeria
SSA	Sub-Saharan Africa
ASIA	Asia
LACA	Latin and Central America

Note: ICES = Intertemporal Computable Equilibrium System.

Nigeria GDP growth through 2025 is consistent with the projections proposed by the prudential interpretation of Nigeria Vision 20: 2020. This prudential interpretation basically shifts to 2025 the Nigeria Vision targets for 2020. Between 2025 and 2050, GDP growth rates are assumed to remain positive but to decline. Specifically (figure I.2) annual real GDP growth for Nigeria is assumed to peak at about 10 percent in 2016–17, decline to 8 percent in 2025, and eventually hit 4 percent in 2050. Thus, average annual growth rates are 9 percent for 2010–25 and 5.7 percent for 2025–50.

Another characteristic of the baseline construction is the sectoral composition of Nigerian value-added. Vision 20: 2020 assumes the following ranges for sectoral value-added: services, 45–75 percent; manufacturing, 15–30 percent; and agriculture, 3–15 percent.

These estimates were revised for purposes of this study, in particular increasing the share attributed to agriculture. In fact, the expected 15 percent still seems too low given the production and productivity growth rate targets for the sector (tripling domestic agricultural productivity by 2015 and doubling that again by 2020). There are additional assumptions for agriculture: the fast rise in productivity is to be supported by an increase in irrigated land from 1 percent in 2010 (Vision 2020) to 20 percent in 2020.

More realistic assumptions estimate that by 2025 (rather than 2020) agriculture's value-added to total GDP will be about 21 percent (slightly less than half the 42 percent in 2010), with manufacturing contributing about 18 percent,

Figure I.2 Annual GDP Growth for Nigeria

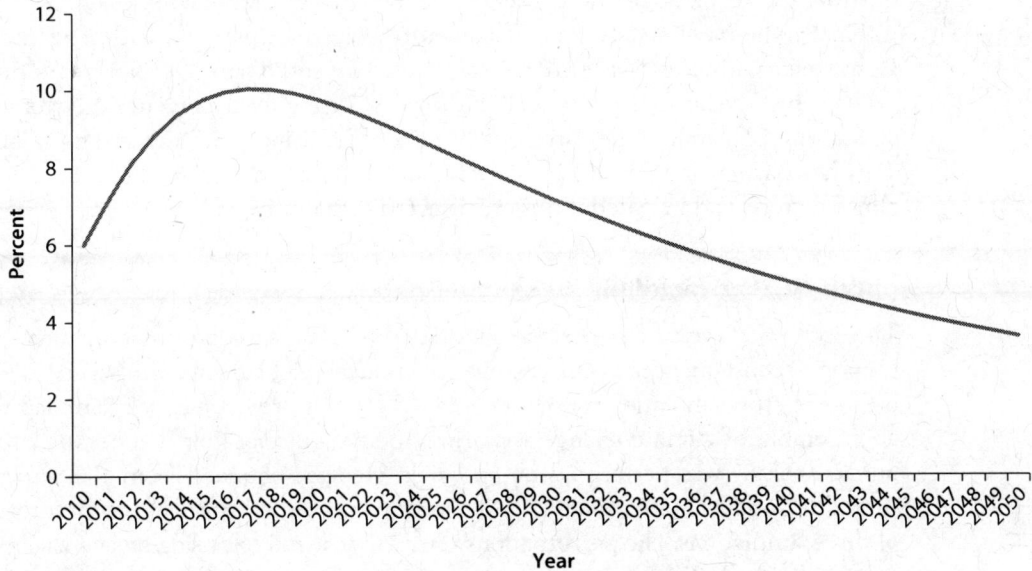

Source: Authors' calculations based on data sources listed in table 3A.1.

Figure I.3 Evolution of Nigerian Value Added by Macrosector

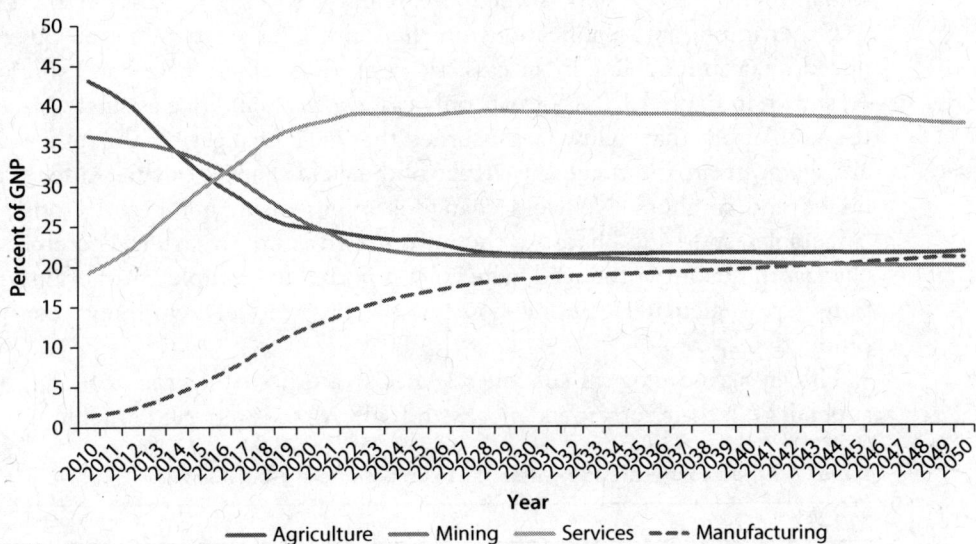

Source: Authors' calculations based on data sources listed in table 3A.1.

mining about 17 percent, and services close to 40 percent. The ICES baseline replicates these targets well (figure I.3).

After 2025 it is assumed that the sectoral shares stabilize except for a slight increase of manufacturing, at the expense of mining, to 21 percent in 2050.

The ICES baseline also covers the targeted agricultural productivity increases, but they are treated as exogenous scenario data: it is simply assumed that Nigeria will be able to reach its targets without any specific assumptions that qualify or quantify the technologies and costs necessary to meet these increases.

Finally, the baseline also contains assumptions about progress with irrigation. Consistent with the Master Plan for Irrigation and Dam Development but delayed by 5 years, it is expected that in 2025 roughly 5 percent of Nigerian agriculture (2.1 million hectares) will be irrigated, rising to 25 percent by 2050. This has particular relevance when climate change impacts are computed for rain-fed crops and irrigated crops are affected differently.

Inputs to the ICES Model and Simulations

The economic assessment, performed for 2010–50, aims to determine the macroeconomic consequences of climate change–induced yield changes in Nigeria. The output of crop simulation models developed for this project has been applied to an ensemble of climate change scenarios. To save computational time, the economic analysis was performed on the yield changes obtained from the unperturbed regional climate model (RCM) and two perturbed simulations for a total of three simulations. The perturbations refer to those models using on average less pessimistic and more pessimistic impacts on crops yields.[2] Specifically the National Center for Atmospheric Research (NCAR) implies moderate yield losses and Geophysical Fluid Dynamics Laboratory (GFDL) is more pessimistic. These scenarios plus the two unperturbed ones span the space of all possible solutions.

Other important qualifications on the input data are that the output produced by crop modeling is for cassava, yams, rice, millet, maize, and sorghum. As shown in figure I.1 ICES treats only cassava, yam, and rice as single crops. In the simulations that follow it is assumed that yield changes for the "other cereals" aggregate are the weighted average of the yield changes calculated for maize, millet, and sorghum. No yield change is imposed on "non-cereal crops" and "vegetables and fruits" because the information is not available. Therefore the changes in quantity produced, which are reported for completeness of information, are calculated endogenously by the model assuming constant productivity.

Fifteen agro-ecological subzones (AESZs) are used for crop modeling (see appendix A); their correspondence with ICES AEZs is reported in table I.2.

Table I.2 Agro-ecological Zoning Relation to Crop Modeling and ICES

ICES-AEZ	NSPFS[a]
AEZ 1	AESZ 7
AEZ 2	AESZ 1 and AESZ 2
AEZ 3	AESZ 9 and AESZ 5
AEZ 4	AESZs 10, 8, 11, 4, 3
AEZ 5	AESZs 14, 12, 13, 6
AEZ 6	AESZ 15

Note: AESZ = agro-ecological subzone; AEZ = agro-ecological zone; ICES = Intertemporal Computable Equilibrium System; NSPFS = National Special Programme for Food Security.
a. See appendix A, as used in Mereu and Spano 2011.

Table I.3 Inputs to the ICES Model: Changes in Irrigated and Rain-Fed Crop Yields from Current Climate

percent

		Rice	Cereal crops	Cassava	Yams
RCM constant CO$_2$					
2020	AEZ1	0.0	0.0	0.0	0.0
	AEZ2	−3.7	−4.3	0.0	0.0
	AEZ3	−3.4	−6.3	−3.4	−0.8
	AEZ4	−6.1	−6.2	−3.1	−0.7
	AEZ5	−5.8	−5.4	−6.4	−1.4
	AEZ6	−2.4	−7.2	−9.0	−0.9
2050	AEZ1	0.0	0.0	0.0	0.0
	AEZ2	−25.3	−18.8	0.0	0.0
	AEZ3	−13.0	−12.7	−12.4	−9.3
	AEZ4	−19.3	−16.4	−17.6	−14.5
	AEZ5	−18.6	−13.9	−18.4	−13.3
	AEZ6	−15.5	−12.2	−21.3	−14.0
NCAR constant CO$_2$					
2020	AEZ1	0.0	0.0	0.0	0.0
	AEZ2	1.5	7.6	0.0	0.0
	AEZ3	−13.3	−12.2	4.3	4.1
	AEZ4	−8.1	−6.2	6.2	5.6
	AEZ5	−9.6	−3.7	15.7	7.4
	AEZ6	−7.4	−7.7	7.0	3.4
2050	AEZ1	0.0	0.0	0.0	0.0
	AEZ2	−19.1	−2.8	0.0	0.0
	AEZ3	−34.3	−14.9	−3.3	−0.8
	AEZ4	−17.9	−11.8	−5.2	−2.4
	AEZ5	−27.5	−9.0	1.8	−4.2
	AEZ6	−23.0	−11.9	−4.9	−10.0
GFDL constant CO$_2$					
2020	AEZ1	0.0	0.0	0.0	0.0
	AEZ2	−0.3	2.5	0.0	0.0
	AEZ3	−15.2	−9.7	5.6	2.8
	AEZ4	−8.2	−4.6	5.3	2.9
	AEZ5	−9.0	−5.9	1.0	0.4
	AEZ6	−6.1	−10.2	−8.9	−1.2
2050	AEZ1	0.0	0.0	0.0	0.0
	AEZ2	−25.2	−17.4	0.0	0.0
	AEZ3	−42.0	−16.5	−5.5	−4.3
	AEZ4	−24.4	−13.9	−10.8	−7.5
	AEZ5	−27.3	−12.2	−13.4	−10.6
	AEZ6	−20.1	−11.8	−21.3	−15.2

Note: AEZ = agro-ecological zone; GFDL = Geophysical Fluid Dynamics Laboratory; ICES = Intertemporal Computable Equilibrium System; NCAR = National Center for Atmospheric Research; RCM = Regional Climate Model.

The inputs for all the ICES simulations are reported in table I.3. They are changes in net yields and account for the differences in how climate change affects rain-fed and irrigated crops.

Notes

1. The agriculture components are livestock and fishery.
2. The disclaimer "on average" is needed because there is no perturbed simulation providing the best or worst yield outcome in all AEZs and for all crops.

References

Armington, P. 1969. "A Theory of Demand for Products Distinguished by Place of Production." *International Monetary Fund Staff Papers* 16: 159–78.

Avetisyan, M., U. Baldos, and T. Hertel. 2011. "Development of the GTAP Version 7 Land Use Data Base." GTAP Research Memorandum 19, Center for Global Trade Analysis, Purdue University, West Lafayette, IN.

Lee, H.-L., T. W. Hertel, B. Sohngen, and N. Ramankutty. 2005. "Towards an Integrated Land Use Database for Assessing the Potential for Greenhouse Gas Mitigation." GTAP Technical Paper 25, Center for Global Trade Analysis, Purdue University, West Lafayette, IN. https://www.gtap.agecon.purdue.edu/resources/res_display.asp? RecordID=1900.

Mereu, V., and D. Spano. 2011. "Climate Change Impacts on Crop Production." Report for the World Bank tender, "Climate Risk Analysis over Nigeria."

Narayanan, G. B., and T. L. Walmsley, eds. 2008. "Global Trade, Assistance, and Production: The GTAP 7 Data Base." Center for Global Trade Analysis, Purdue University, West Lafayette, IN.

Nwafor, M., X. Diao, and V. Alpuerto. 2010. *A 2006 Social Accounting Matrix for Nigeria: Methodology and Results*. Nigeria Strategy Support Program (NSSP), Report NSSP007, International Food Policy Research Institute (IFPRI), Washington, DC.

Robust Decision Making in Irrigation

Climate Change and Robust Decision Making

As discussed in the main text, Nigeria's climate is likely to become warmer. Precipitation and its pattern (number of wet days and duration of dry periods) are also projected to change, with some areas becoming wetter, others dryer. As a result, runoff, for instance, is likely to be affected by climate change, but uncertainty between regions and between models is large.

Reservoir size is typically designed to ensure sufficient storage to provide a set continuous or seasonal flow based on past weather patterns, but historical data may no longer be adequate to guide long-term investment. Climate change impact needs to be considered in the design of new water mobilization and irrigation scheme projects. With climate change, a given storage identified based on historical data can receive less or more water than expected and produce less or more benefit. The design can be adapted to a given future climate at a certain cost, the adaptation cost, which is the extra capital cost of building storage or irrigated area (which can be negative if less storage or area is needed than for the historical climate). The benefit is the extra revenue from selling irrigated crops (which can also become negative if too little storage is built and fewer crops produced than in the reference).

Block and Brown (2008) proposed an approach for integrating climate change information into evaluation of hydropower dam projects to address climate risk management decisions. Their approach estimates climate change impact for a set of scenarios by calculating how benefit-cost ratios change for each precipitation and temperature combination. Jeuland (2010) also developed a sophisticated hydroeconomic model for integrating climate change impacts into planning for water resources infrastructure. Although these approaches give a sense of the influence of climate change on design options, the analysis is based on a choice of alternatives that may not include options that minimize the economic impact. These approaches need to be combined with other decision tools to help the planner optimize the infrastructure design.

Planning, when climate change is uncertain, involves reducing regrets over the possible future climates. Different strategies have been proposed to decide despite uncertainty (Hallegate 2009; Lempert and Kalra 2011). A particularly promising approach (Defra 2011; Dessai and Wilby 2010) is to identify "robust decisions"—those that perform better than the alternatives over a wide range of plausible futures, even if they are not the best possible approach for any one view of the future. Among them, appealing strategies are "no regret" (adaptive measures whose socioeconomic benefits exceed their costs no matter the extent of climate change) or "low regret" (adaptive measures for which the costs are relatively low and the benefits, although primarily realized under projected future climate change, may be relatively large), because they are directed at maximizing the return on investment when certainty of the associated risk is low (Hills and Benett 2010). In this book, the robust decision is identified by minimizing regrets for different climate scenarios.

Methodology

The approach adopted for this book consists of identifying an option that minimizes regrets for a range of possible future climate scenarios. The regrets are defined as the difference in economic return between the chosen option ("no foresight") and the best possible option calculated for each scenario ("perfect foresight"). Net present value (NPV) is used to estimate the economic return. Two objective functions were minimized: the average and the maximum regrets among scenarios, each reflecting a different degree of risk aversion. The optimizations were carried out with respect to one of two decision variables, the amount of stored water or the irrigated area.

Estimation of Cost and Revenue

The cost estimate of an irrigation dam has two components: (1) the storage cost, which includes the capital cost for the dam and the operation and maintenance (O&M) cost; and (2) the irrigation infrastructure cost, which also includes capital and O&M costs. In this section, multipurpose dams are not considered, so the benefit comes only from the value-added by irrigated crop production over rain-fed agriculture.

The cost-benefit analysis presented here is intended not to inform project appraisal, which would require a richer dataset, but to estimate the order of magnitude of the NPV with and without perfect foresight, in order to test how the methodology could apply. A sensitivity of the model to the main parameters was tested out and the results are presented in a later subsection.

Storage Cost

The storage cost is estimated using the method described in Ward *et al.* (2010). Relationships between slope and construction costs per cubic meter are developed for 11 reservoir size classes based on U.S. data. The capital cost per unit of water stored is a parabolic function of the slope, and the parameters of the equation depend on the storage range considered. Slopes are obtained from ArcSWAT (the hydrological model used in this study). The O&M cost is estimated at 2 percent of the capital cost (Ward *et al.* 2010), and African costs are assumed to be 68.5 percent of U.S. costs (Kirshen 2007).

Irrigation Infrastructure Cost

The cost of irrigation infrastructure is assumed to be US\$6,000/hectare (ha) for the capital cost with US\$20/ha for O&M. This order of magnitude can be found in studies for large- and medium-scale Sub-Saharan African schemes in 2004 (World Bank 2007; You *et al.* 2009), but there is considerable variability among irrigation schemes. In Nigeria, the cost of large-scale irrigation projects built in the 1990s varied (JICA 1995) from US\$3,700/ha (Balanga, 1,800 ha under gravity irrigation) to US\$24,500/ha (Bakalori, 23,000 ha also under gravity irrigation).[1] For consistency with other data sources (production price, yield), 2004 costs were used, but it should be noted that irrigation cost has greatly increased since 2004.

Revenue from Irrigation

It is assumed that irrigation development would convert rain-fed into irrigated areas. The net revenue of conversion is therefore the difference between revenue from irrigation and revenue from rain-fed agriculture.

Revenue from irrigated and rain-fed crops is found by the following equation:

$$R_i = (Y_i P_i - C_i) A_i \qquad (J.1)$$

where R_i is the revenue of the crop i, Y_i yield, P_i price, C_i the production cost per ha, and A_i the production area of crop i.

The cropping pattern is assumed to be adapted to the climate of each hydrological area (HA). The crop pattern used for irrigated crops is given in table J.1 (JICA 1995). Under these assumptions, rice is the predominant irrigated crop in the south of Nigeria and maize or other cereal (wheat) is grown in the north and central zones, with a vegetable as a second crop.[2]

The corresponding water requirement (defined here as the water to be diverted to irrigate) is also shown in table J.1. It is calculated using a Penman equation for evapotranspiration, assuming effective rainfall of 70 percent of the average and 50 percent of water losses in the supply network.

The yield and the production cost (input, labor) observed in 2004 in the public irrigation schemes were used for crop data (FAO 2004). Yield of irrigated crops is available for several irrigation schemes throughout the country and the cost of production corresponds to production in the Kano Irrigation Project. For production costs in other parts of Nigeria it was assumed that the ratio production cost/benefits (yield time price) was constant for irrigated schemes. Assumptions on yield are conservative because in the future higher input and better irrigation supply may increase yields, and revenues.

Table J.1 Intensity of Irrigated Crop Production and Corresponding Water Requirement by Hydrological Area

| Hydrological area | Crop intensity (%) | | | | Total water requirement (m³/ha) |
| | Rice | | Other cereals and vegetables | | |
	Wet	Dry	Wet	Dry	
HA1	10	0	90	50	9,600
HA2	30	25	70	55	11,400
HA3	30	25	70	55	13,400
HA4	30	25	70	55	11,600
HA5	90	70	10	10	13,800
HA6	90	70	10	10	14,600
HA7	90	70	10	10	13,200
HA8	10	0	90	50	9,600

Source: JICA 1995.

Data for rain-fed agriculture cropping patterns and yield were extracted from the database collected for the Africa Infrastructure Country Diagnostic.[3] State-level data are available. As suggested by You *et al.* (2009), the production cost was estimated at 67 percent of the benefits (yield × price). This value falls within the range of values for irrigated production (53–77 percent of the benefits).

Crop price corresponds to the 2004 producer price from the Food and Agriculture Organization of the United Nations (FAO).[4] The FAO database provides a national average for a given year, but prices are highly variable, both temporally and spatially, depending on product availability (prices at harvest are lower) and proximity to the market (prices are higher when crops are produced close to markets). For example, the survey done for review of the public irrigation sector in Kano in 2004 showed that prices at harvest were much lower than the national average (40 percent lower for cereals and up to 80 percent for vegetables). It is assumed that the market condition will improve when planning irrigation schemes. At the planning stage, assumptions about prices should reflect local conditions.

Figure J.1 shows revenue per hectare of irrigated and rain-fed agriculture in different HAs. The revenue differences between the two in HA5, HA6, and HA7 are too low to make an investment in an irrigation dam viable. In practice, multipurpose dams are preferred in this area, but they cannot be used for application of this methodology. So for purposes of this study irrigation revenue is artificially increased (double the yield). (This change could also correspond to a change in price or crop pattern.)

Calculation of the NPV and Investment Plan

The NPV was calculated as follows:

$$\text{NPV} = \sum_{t=1}^{T} \left(\frac{1}{1-r} \right)^t \left[\left(R_{\text{irrigation}} - R_{\text{rainfed}} \right) * f_1 - \text{Capital cost} * f_2 - \text{O\&M cost} * f_3 \right]$$

$$(J.2)$$

Figure J.1 Revenue for Irrigated and Rain-Fed Crops for Different HAs

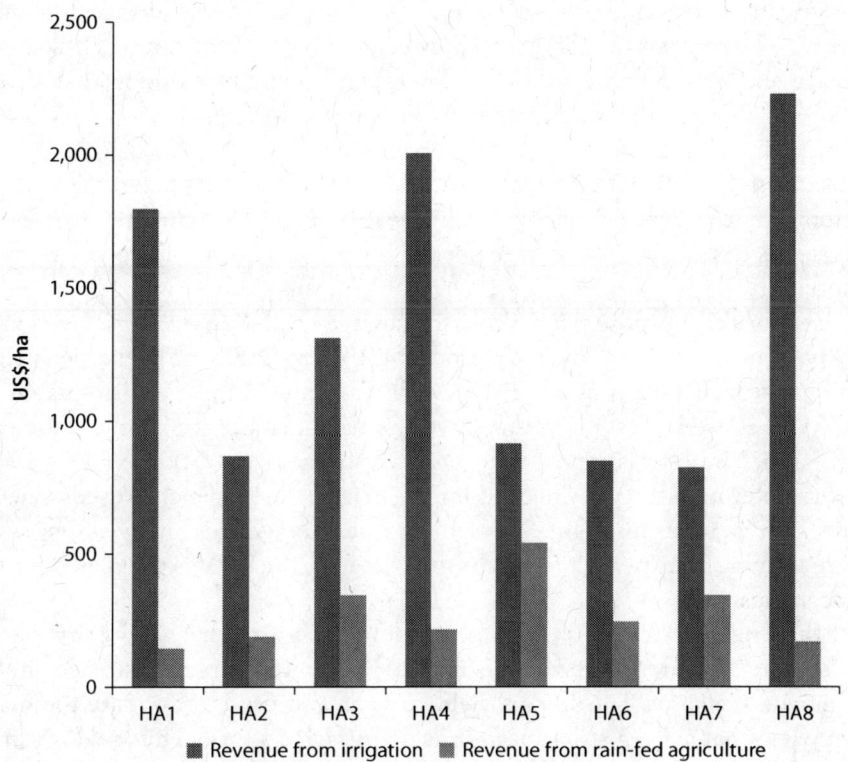

Source: Authors' calculations based on data sources listed in table 3A.1.
Note: HA = hydrological area.

Table J.2 Time Assumptions for Investments and Benefits[a]

Year	F_1 (capital cost)	F_2 (O&M cost)	F_3 (revenue)
1	0.3	0	0
2	0.3	0	0
3	0.3	0	0.3
4	0.1	0.5	0.6
5	0	0.5	1
6...	0	1	1
...50	0	1	1

Note: O&M = operation and maintenance.
a. f_1, f_2, and f_3 are the factors used to calculate net present revenue.

where T is the duration of the project; r the discount rate; R the revenue; f1, f2, and f3 are the time factors specified in table J.2. The investment plan is based on You *et al.* (2009), but with a shorter construction period (four rather than eight years) because the projects being assessed are of medium size (see the following discussion for projects selected). The discount rate used was 13 percent; it was included in the Sensitivity Analysis.

Calculation of the Regrets

The irrigation costs and revenues are used to estimate regrets for each climate scenario. The regret is the difference between the NPV of the option chosen now (storage and irrigation area with no foresight) and the best option that should have been chosen to irrigate the same area (perfect foresight).

Estimating the NPV with Perfect Foresight for Each Climate Scenario

If storage is the decision variable adjusted to account for climate change, the first step is to estimate the storage needed for the yield from the targeted irrigation area in the 11 climate scenarios available in this study. Storage-yield curves are used for this purpose. They show the storage needed to deliver a given yield 100 percent of the time over a period of 60 years (2005–65). The peak algorithm approach (Ward *et al.* 2010) with monthly runoff data simulated by ArcSWAT is used to establish the curves for each climate scenario and for each dam basin. The estimation of the water requirement per hectare (table J.1) is used to determine the yield needed for the irrigated area, which is used to determine the "perfect foresight" storage on the storage-yield curve. Once the optimal storage for "perfect foresight" is estimated, the NPV is calculated for the 11 scenarios.

If the irrigated area is adjusted to account for climate change, the first step is to estimate the area that can be irrigated with the targeted storage in the 11 climate scenarios. The storage yield curve is also used to estimate the maximum yield that a given storage can provide in each scenario. This yield is turned into irrigated area using the water requirement per hectare (water requirement holds constant in the future).[5] NPV is then calculated.

Determining the "No-Foresight" Case That Minimizes Regrets

The "no-foresight" case corresponds to the design alternative chosen now. If storage is the variable that will be modified to minimize regret, for optimization a first storage value is assumed and the corresponding yield in each potential future is estimated using the storage-yield curve. This yield corresponds to a certain irrigated area (water requirement holds constant). If the irrigated area is smaller than the area the project targets, there is a loss of irrigation benefits. It was assumed that the irrigated area could be no larger than the targeted area, maximum of which is fixed by the size of the infrastructure.[6] The economic return (NPV) of the "no-foresight" storage and its corresponding irrigated area is then calculated for each climate scenario and compared to the NPV with "perfect foresight." The difference is the regret. Figure J.2 illustrates the regrets for different values of storage for a given climate scenario. The "no-foresight" storage is then adjusted to minimize the average (or the maximum) regret among possible future climates. An example of the minimization of maximum regrets between a wet and dry future climate is shown in figure J.3. The solver of Excel was for optimization between the 11 climate scenarios.

If irrigated area is the optimized variable, the steps are the same but the irrigated area is used in place of the storage.

Figure J.2 Regrets Calculation for Storage Optimization for a Given Climate Scenario, with Perfect Foresight Storage of 18 Million m³

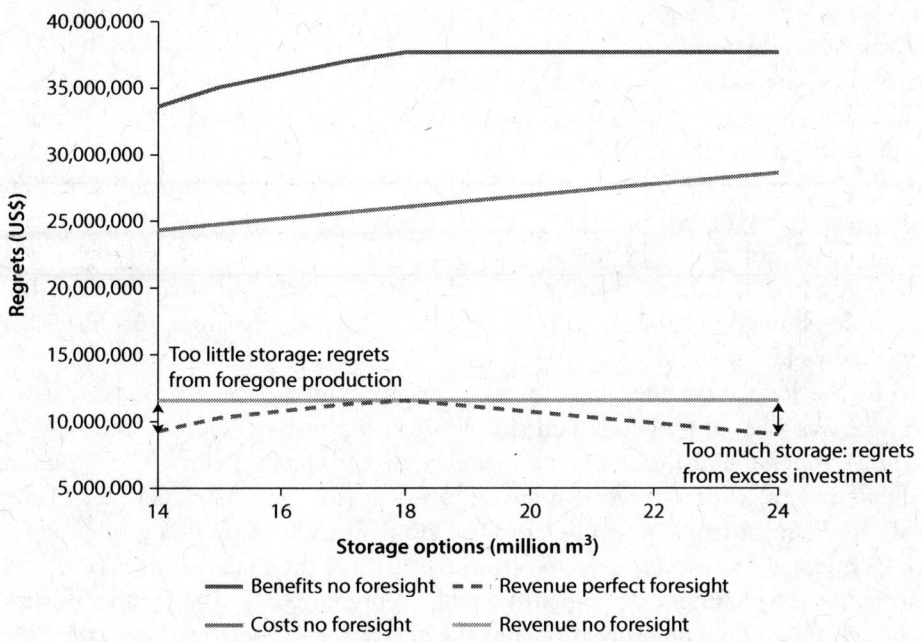

Source: Authors' calculations based on data sources listed in table 3A.1.

Figure J.3 Regrets for Historical Climate and Future Wet and Dry Climates

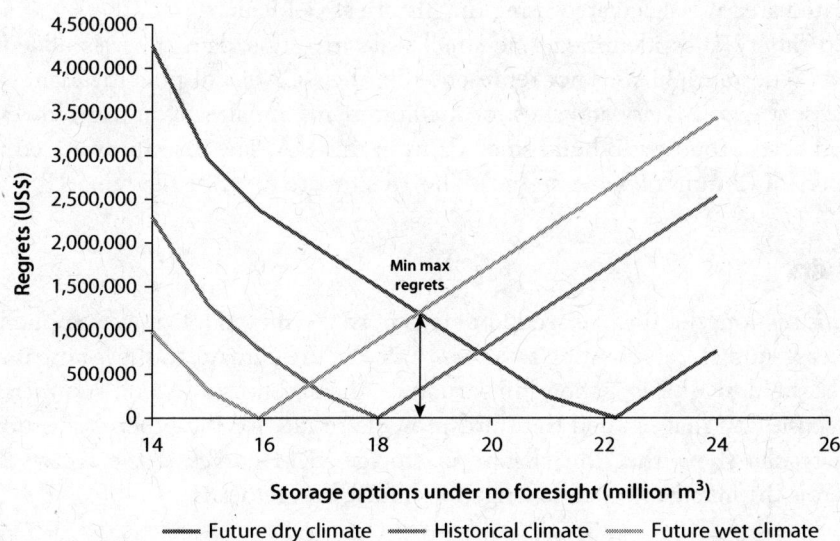

Source: Authors' calculations based on data sources listed in table 3A.1.

The robust decision is the storage that minimizes the maximum (as shown on figure J.3) or the average regrets between climate scenarios.

Dam Site Selection

Based on the federal government's irrigation plans, 18 dam sites were selected. The short-term priority for the government is to rehabilitate and expand existing public irrigation schemes (57 projects). After checking information available for each project, several schemes were identified that will benefit from construction of a dam to expand the area they irrigate. Unfortunately, most will be multipurpose dams and cannot be included in the study. The Zauro polder project was the only irrigation dam identified. For this site, the irrigated area was available and the storage needed to irrigate the area was deduced from the historical storage-yield curve.

In the long term the government plans to construct new irrigation dams. There was a list of potential dam sites in the 1995 water resources study (JICA 1995), which identified 264 medium and large dams (capacity of 20–150 million cubic meters) with a catchment area of 50–500 km^2. This list is not part of the Master Plan for Irrigation and Dam Development (2009–20). Using the following criteria, 17 sites were selected from the list: (1) the main basins where new irrigation development is planned should be represented; (2) the number of sites in each HA should be proportional to the area planned for irrigation in the HA; (3) the catchment size should be larger than 100 km^2 so that the sub-basins used in ArcSWAT are representative of catchment behavior; (4) there is no dam upstream (visible in Google Maps); and (5) dry and wet future climates are represented.[7] The targeted storage was noted in the JICA (1995) document and the irrigated area was calculated using the historical yield curve.

To the 17 sites identified, one small-scale irrigation dam site was added in HA8 (Yedesram). HA8 is not represented in the JICA list of potential dam sites because the area is dry and sites for medium dams are already in use. However, it had been proposed to build small dams in this HA. The case study added is a small dam (2 mm^3 of water stored). The 18 sites are described in table J.3.

Results

In this section, the first subsection details how the methodology was applied in two case studies, representing a wet and a dry future climate, to show how it can affect the design of irrigation infrastructure; the second subsection summarizes the sensitivity analysis; and the third presents results for the other case studies. The results show that the change in storage yield curves is the factor that controls the change needed in irrigation infrastructure design.

A Wet and a Dry Climate Case Study

In this section, wet climates are defined as climates where less storage is needed in the future to provide a given yield compared to the storage needed under the historical climate. In other terms, in a wet climate, the potential irrigated area

Table J.3 Sites Selected for Analysis

Hydrological area	Site	River	State	Catchment area (km²)	Annual runoff (million m³)	Active capacity (million m³)
HA1	Ka	Ka	Sokoto	7,600	608	115
	Zauro	Rima	Kebbi	50,720	5,988	—
HA2	Ajelanwa	Weru	Kwara	690	117	85
	Bakajeba	Jatau	Niger	975	195	147
	Kuda	Kuda	Kaduna	140	34	29
HA3	Kunini	Kunini	Taraba	300	72	23
	Hona Gombi	Dogaba	Adamawa	420	92	61
	Ganye	Ini	Adamawa	660	139	175
	Suntai	Suntai	Taraba	5,200	2,080	50
HA4	Karma	Karma	Nasawara	250	36	144
	Tsorom	Akwenyi	Nasawara	500	110	24
	Ambighir	Ambighir	Benue	200	50	43
	Baushe	Baushe	Plateau	480	96	200
HA5	Oji	Oji	Enugu	310	112	20
HA6	Ibu	Ubi	Ogun	250	104	26
HA7	Ogege	Ogege	Benue	250	93	22
	Moi	Moi	Cross River	150	60	22
HA8	Yedesram	Yedesram	Borno	—	—	2

Note: — = not available.

downstream a given dam will be larger than under the historical climate. (The reverse is true for dry climates: more storage is needed, and potential irrigated area downstream will be smaller.) The historical storage-yield curve is the basis for decisions made that ignore climate change.

The two case studies presented here were selected from the list of 18 sites using the storage-yield curves: the historical storage-yield curve of a wet case is below a majority of future climate curves, and is above it for a dry case. The wet climate corresponds to the Kuda site (HA2) and the dry one to the Karma site (HA4) (see table J.3 for details). The storage yield curves are shown in figure J.4.

Figure J.5 shows the change in mean annual runoff for the two case studies.[8] It is interesting to note that the impact of climate change on mean annual runoff is different from the impact on storage-yield curve. Indeed, the change in storage yield curves reflects the change in runoff variability and more specifically the change in the duration of the dry period. Although longer dry periods are often observed when mean annual runoff decreases, it was observed that an increase in mean annual runoff does not guarantee an increase in the yield delivered by a dam. Deeper analysis of each dam site is needed to confirm this result, because determination of storage-yield curves is based on whether the hydrological model can correctly simulate dry conditions. In this case, ArcSWAT was calibrated against limited numbers of gauging stations and of years. The actual variability of the runoff may not be properly reproduced.

Figure J.4 Storage Yield Curves for 2005–65

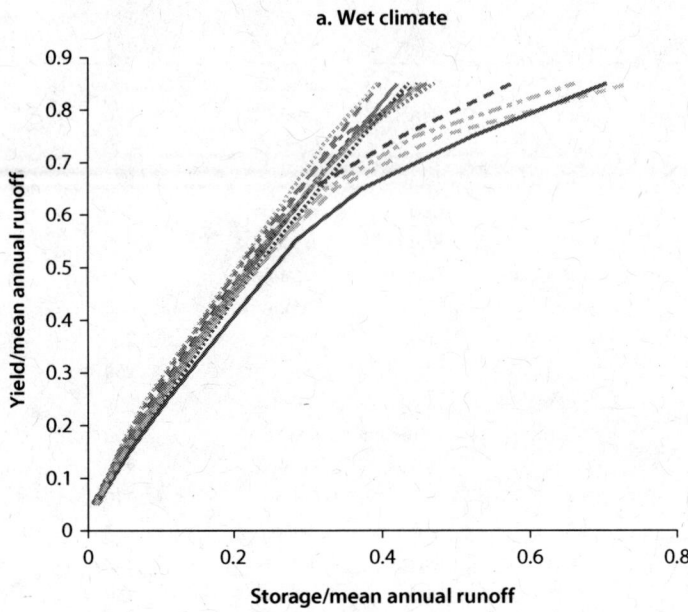

a. Wet climate

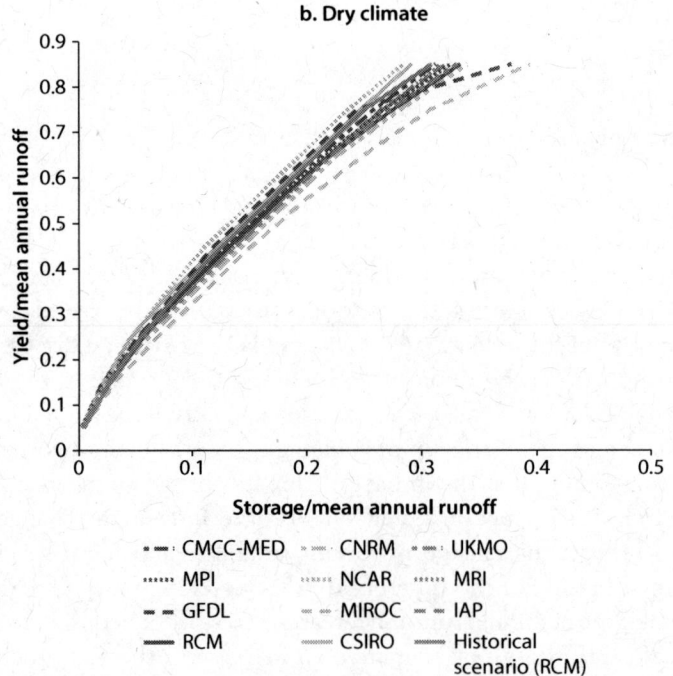

b. Dry climate

: CMCC-MED	CNRM	UKMO
MPI	NCAR	MRI
GFDL	MIROC	IAP
RCM	CSIRO	Historical scenario (RCM)

Source: Authors' calculations based on data sources listed in table 3A.1.
Note: RCM = Regional Climate Model; CMCC-MED = Euro-Mediterranean Center on Climate Change; CNRM = Centre National de Recherches Météorologiques; CSIRO = Commonwealth Scientific and Industrial Research Organization; GFDL = Geophysical Fluid Dynamics Laboratory; IAP = Institute of Atmospheric Physics; MIROC = Center for Climate System Research; MPI = Max Planck Institute; MRI = Meteorological Research Institute; NCAR = National Center for Atmospheric Research; UKMO = United Kingdom Meteorological Office.

Figure J.5 Change in Mean Annual Runoff for the Wet Climate and Dry Climate Cases

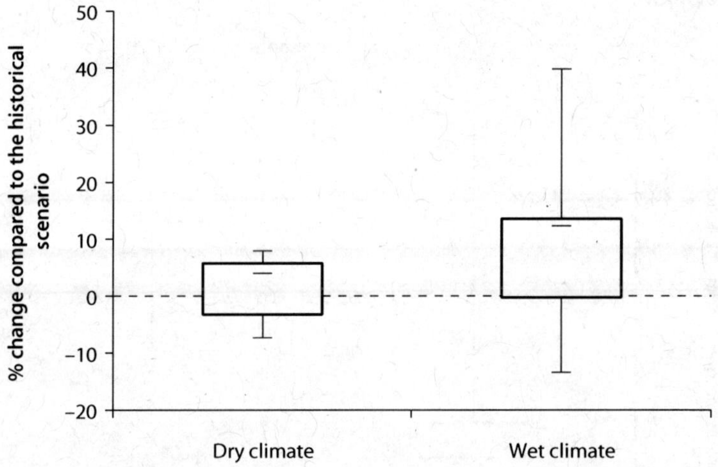

Source: Authors' calculations based on data sources listed in table 3A.1.

The change in design needed to adapt to climate change with perfect foresight in each scenario is shown in figure J.6. The magnitude of changes is linked to the relative position of the historical storage-yield curve compared to the curves projected. The figure shows that the changes required for the wet climate are greater than for the dry climate. Indeed, for the wet climate, the distance between the historical curve and the majority of other curves is important for high values of the ratio yield/mean annual runoff (about 0.7 in this case study). The corresponding decrease in storage, or increase in irrigated area, is therefore relatively high. On the contrary, for the dry climate, the historical curve remains close to the others for the value of the yield/mean annual runoff of this case study (about 0.8). The corresponding change in storage or irrigated area is lower than in the wet climate case.

The NPV of the investment with perfect foresight is calculated with a discount rate of 13 percent. The distribution of NPV is similar to that of the change in storage and in irrigated area.

Figure J.7 shows the distribution of regrets among climate scenarios for different values of storage and irrigated areas chosen with no foresight. The regrets (difference between the NPVs with no foresight and with perfect foresight) are expressed as a percent of project capital costs (irrigation and storage capital costs).

For the wet climate case, the value of storage that minimizes average regrets (12.65 mm^3) and the value that minimizes maximum regrets (12.63 mm^3) are about 26 percent lower than the storage estimated under the historical conditions (17 mm^3). This is consistent with the general trend of a future decrease in dry periods observed on the storage yield curves of this case study. Similarly, the values of irrigated area that minimize average and minimum regrets are 20 and 5 percent higher than the irrigated area chosen with historical data.

Figure J.6 Change in Storage and Irrigated Area Needed to Adapt to Climate Change with Perfect Foresight

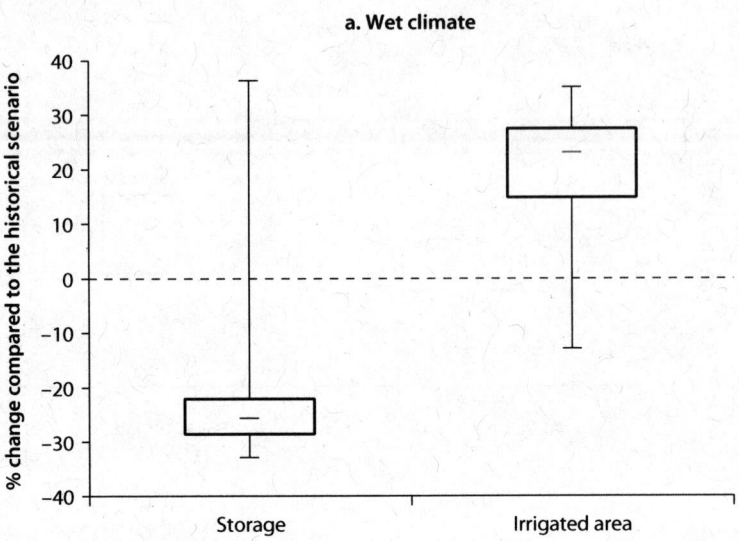

a. Wet climate

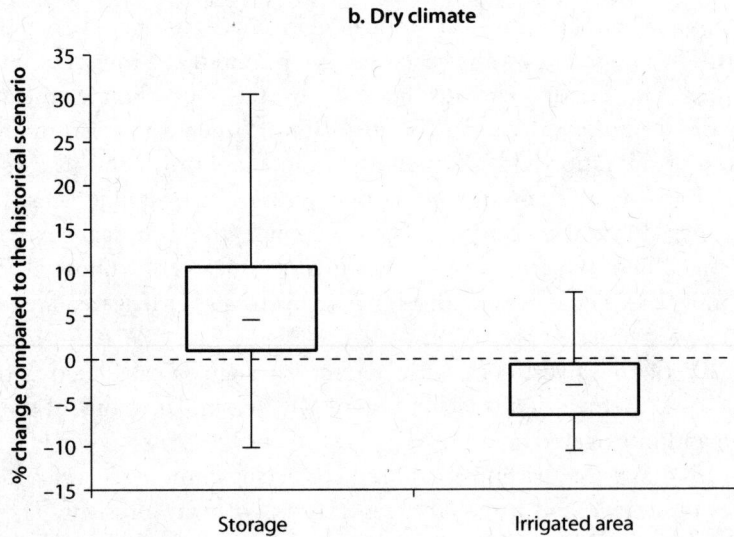

b. Dry climate

Source: Authors' calculations based on data sources listed in table 3A.1.

For the dry climate case, to minimize average regrets storage should be 6 percent higher and irrigated area 3 percent lower than the historical design; to minimize maximum regrets storage should be 7 percent higher or irrigated area 1 percent smaller. These changes are not as significant as the ones calculated for the wet climate case. Indeed, in the latter, the values of storage and irrigated area with perfect foresight are significantly different from the historical ones (see the results section), which is not true for the dry climate case.

Figure J.7 Regrets Distribution for Design Options with No Foresight for the Wet and Dry Climate Cases, Storage Optimized and Irrigated Area Optimized

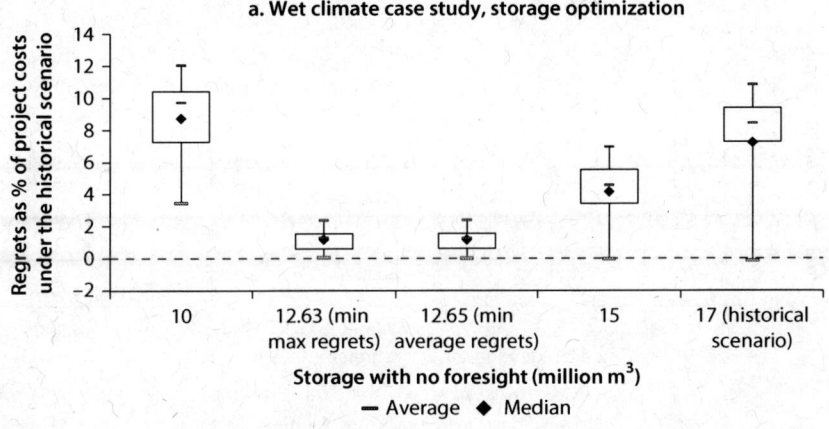

a. Wet climate case study, storage optimization

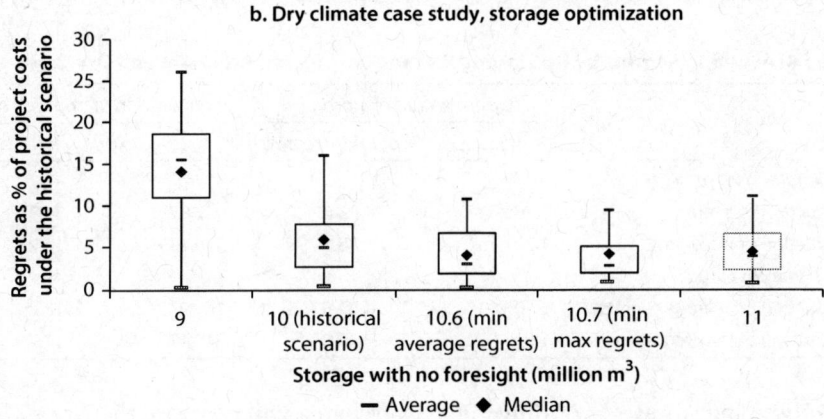

b. Dry climate case study, storage optimization

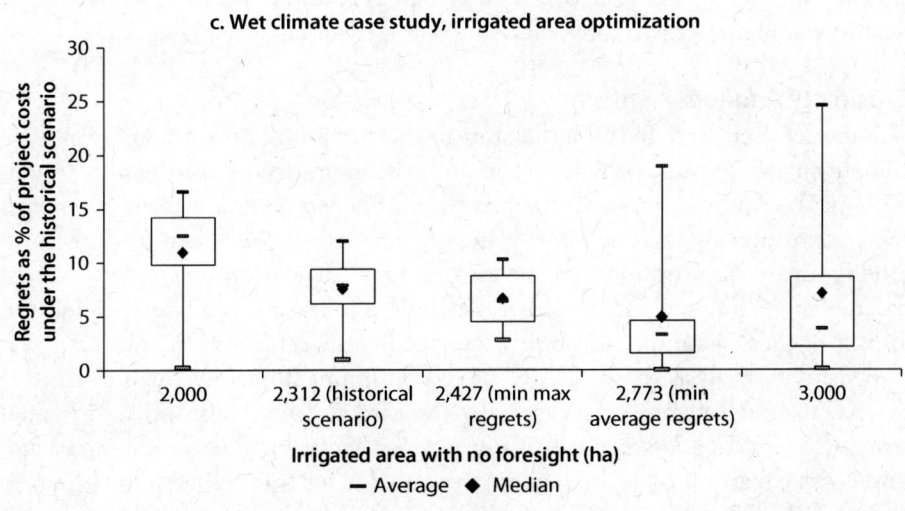

c. Wet climate case study, irrigated area optimization

figure continues next page

Figure J.7 Regrets Distribution for Design Options with No Foresight for the Wet and Dry Climate Cases, Storage Optimized and Irrigated Area Optimized *(continued)*

d. Dry climate case study, irrigated area optimization

Source: Authors' calculations based on data sources listed in table 3A.1.

Table J.4 Avoiding Regrets by Optimizing Storage or Irrigated Area, Wet and Dry Cases

	Optimization of storage		Optimization of irrigated area	
	US$	% of initial regrets	US$	% of initial regrets
Wet climate case study				
Reduction of average regrets	1,266,648	83.8	547,829	34.7
Reduction of max regrets	1,771,162	78.2	356,789	14.2
Dry climate case study				
Reduction of average regrets	309,844	32.8	169,932	18.2
Reduction of max regrets	1,077,600	40.8	240,471	11.7

Optimizing storage or optimizing irrigated area will decrease the regrets. The gain in NPV, or regrets avoided, can be as high as 83.8 percent of the initial regret for the wet climate case study with an optimized storage (see table J.4).

Sensitivity Analysis Results

The model used to determine the storage that minimizes the regrets related to climate change is based on calculation of the economic return of the investment, NPV in this study. All the parameters that influence NPV are likely to change the distribution of regrets. More generally, changes in NPV are also likely to modify the design options: many assumptions have been made to estimate parameter values, and the sensitivity analysis is a means of understanding the impact of these assumptions on projected economic return on the project.

Sensitivity analysis shows that NPV and optimum storage are highly sensitive to several model parameters, especially the discount rate, crop prices, irrigated crop yield, and to a lesser extent water use per hectare and irrigation infrastructure costs. Clearly, if costs are higher than benefits for the storage values, there is no optimal storage that minimizes regrets (or storage equals zero).

Results for All Case Studies

Tables J.5 and J.6 show the main results obtained by minimizing the maximum regrets and the average regrets. The magnitude of design change required to adapt to climate change and the regrets avoided are discussed in the text. The difference between the functions to be minimized (maximum or average regrets) is discussed next—the magnitude of design change is not the same for both types of optimization. For example, the change in storage needed to minimize maximum regrets may be lower (Zauro case study) or higher (Kunini case study) than for minimization of average regrets. The change required for the two optimizations can even be opposite compared to the baseline (see the storage of the Ka case).

Discussion

How to Use the Methodology

The proposed method is a simple tool that quantifies and minimizes regrets in designing an irrigation dam. Several observations can be drawn from the case studies.

First, adapting irrigation dam design to climate change requires a focus on the projected change in storage-yield curves. Mean annual runoff, although relatively easy to estimate, should be considered as just a first approximation; to properly estimate future storage-yield curves, a more comprehensive hydrological analysis would be required to correctly capture (1) the beginning and the end of the dry periods, and (2) the volume of runoff during the wet season. These two parameters should be specifically examined while calibrating and validating the hydrological model.

Second, other factors can be as important as climate in determining robust investment decisions, as confirmed by sensitivity analysis on the discount rate, irrigated crop yield, crop price, water use per hectare, and storage cost. In addition, simulated runoff was calibrated against a limited number of stations and is also a source of uncertainty, although that is not quantified here. Additional analysis that takes into account the possible range of local parameter values would be needed to address this issue. Furthermore, these investments should be part of a basin-wide water management plan, taking into account other water users (including environmental flows) and the increasing demands that will result from climate change.

In terms of decision rule, this analysis considered minimization of maximum regrets and of average regrets, which led to different results. Minimizing maximum regrets might involve looking at an outlier scenario where regrets are very high, in which case the optimum design will be different from the one found by minimizing average regrets. Using maximum regrets is a conservative option that can be chosen if the decision maker has equal confidence in all climate projections. Minimizing average regrets may be preferred, depending on the decision maker's tolerance for risk.

Table J.5 Minimizing Maximum Regrets: Results for All Case Studies

Hydrological area	Site	State	Historical scenario (baseline)			Minimizing maximum regrets (optimization of storage)				Minimizing maximum regrets (optimization of irrigated area)			
			Investment cost (US$)	Storage (mm³)	Irrigated area (ha)	Initial regrets as % of investment cost	Optimized storage (mm³)	% change in storage	Avoided regrets (% of initial regrets)	Initial regrets as % of investment cost	Optimized irrigated area (ha)	% change in irrigated area	Avoided regrets (% of initial regrets)
HA1	Ka	Sokoto	136,292,084	92	19,169	4.8	92	0.2	4.0	86.2	22,880	19.4	66.2
	Zauro	Kebbi	78,788,952	59	10,572	9.9	56	−5.8	15.9	68.8	14,503	37.2	39.4
HA2	Ajelanwa	Kwara	54,519,139	68	6,119	18.0	34	−49.8	73.4	48.6	10,309	68.5	48.5
	Bakajeba	Niger	93,400,987	103	11,611	12.1	78	−24.4	50.1	35.2	15,612	34.5	37.5
	Kuda	Kaduna	20,928,066	17	2,312	10.8	13	−25.7	78.2	12.0	2,428	5.0	14.2
HA3	Kunini	Taraba	23,286,589	18	2,204	22.7	10	−42.3	78.8	39.2	2,945	33.6	41.9
	Hong Gombi	Adamawa	39,018,695	49	4,168	15.0	32	−35.2	55.1	82.2	7,034	68.8	54.3
	Ganye	Adamawa	93,319,881	110	9,807	15.7	105	−4.7	10.8	22.2	10,000	2.0	4.3
	Suntai	Taraba	105,968,152	40	15,312	6.2	41	2.3	35.7	34.5	18,467	20.6	44.6
HA4	Karma	Nasawara	16,533,659	10	2,474	16.0	11	7.2	40.8	12.4	2,443	−1.2	11.7
	Tsorom	Nasawara	32,791,447	19	4,185	8.3	19	2.2	18.8	18.1	4,557	8.9	34.6
	Ambighir	Benue	29,985,029	30	3,487	12.3	28	−5.2	12.5	24.1	3,810	9.2	22.2
	Baushe	Plateau	35,390,849	44	3,705	21.1	26	−40.6	69.9	58.2	5,573	50.4	50.1
HA5	Oji	Enugu	31,330,859	18	3,868	8.2	16	−8.3	25.6	8.3	4,086	5.6	19.9
HA6	Ibu	Ogun	33,258,245	25	3,814	15.4	21	−17.5	17.2	5.7	3,778	−0.9	4.8
HA7	Ogege	Benue	31,512,048	22	3,725	8.2	21	−3.0	10.2	8.8	3,786	1.6	6.3
HA8	Moi	Cross River	30,928,473	22	3,466	13.2	17	−23.5	56.8	12.2	3,956	14.1	39.3
	Yedesram	Borno	3,733,548	5	298	41.2	2	−55.3	68.5	181.8	835	180.5	54.8

Source: Authors' calculations based on data sources listed in table 3A.1.

Table J.6 Minimizing Average Regrets: Results for All Case Studies

Hydrological area	Site	State	Historical scenario (baseline)			Minimizing average regrets (optimization of storage)				Minimizing average regrets (optimization of irrigated area)			
			Investment cost (US$)	Storage (mm³)	Irrigated area (ha)	Initial regrets as % of investment cost	Optimized storage (mm³)	% change in storage	Avoided regrets (% of initial regrets)	Initial regrets as % of investment cost	Optimized irrigated area (ha)	% change in irrigated area	Avoided regrets (% initial regrets)
HA1	Ka	Sokoto	136,292,084	92	19,169	3.0	78	−15.0	23.2	49.6	23,986	25.1	85.5
	Zauro	Kebbi	78,788,952	59	10,572	5.3	50	−14.6	13.0	31.6	13,902	31.5	39.8
HA2	Ajelanwa	Kwara	54,519,139	68	6,119	15.4	29	−57.2	91.0	31.6	11,740	91.9	68.7
	Bakajeba	Niger	93,400,987	103	11,611	9.6	52	−49.6	85.3	22.6	19,146	64.9	60.5
	Kuda	Kaduna	20,928,066	17	2,312	7.2	13	−25.6	83.8	7.5	2,773	19.9	34.7
HA3	Kunini	Taraba	23,286,589	18	2,204	17.4	11	−40.7	83.5	27.1	3,318	50.6	70.0
	Hong Gombi	Adamawa	39,018,695	49	4,168	9.3	25	−49.0	117.8	55.1	7,763	86.3	73.3
	Ganye	Adamawa	93,319,881	110	9,807	12.6	63	−43.0	73.6	13.5	11,815	20.5	51.1
	Suntai	Taraba	105,968,152	40	15,312	2.6	41	2.7	8.0	11.5	17,113	11.8	33.1
HA4	Karma	Nasawara	16,533,659	10	2,474	5.7	11	5.9	32.8	5.6	2,399	−3.0	18.2
	Tsorom	Nasawara	32,791,447	19	4,185	3.6	18	−4.1	5.3	8.1	4,544	8.6	36.5
	Ambighir	Benue	29,985,029	30	3,487	5.8	18	−40.3	32.9	16.3	4,445	27.5	66.8
	Baushe	Plateau	35,390,849	44	3,705	17.0	17	−61.2	79.9	32.2	5,735	54.8	66.1
HA5	Oji	Enugu	31,330,859	18	3,868	3.2	16	−12.0	23.5	3.1	3,970	2.6	15.1
HA6	Ibu	Ogun	33,258,245	25	3,814	6.6	22	−11.7	26.5	3.1	4,210	10.4	18.8
HA7	Ogege	Benue	31,512,048	22	3,725	4.4	19	−13.1	30.1	5.3	4,068	9.2	13.4
	Moi	Cross River	30,928,473	22	3,466	5.7	18	−17.1	71.8	5.1	3,948	13.9	47.3
HA8	Yedesram	Borno	3,733,548	5	298	33.1	2	−55.3	83.7	114.0	933	213.5	73.2

Source: Authors' calculations based on data sources listed in table 3A.1.

Lastly, there are opportunities to further reduce regrets by making the system more adaptable. The method shows how to optimize design variables that are difficult to change over time, but other variables, such as crop pattern or change in water use per hectare can be adapted to the water available in a given year to cope with the inter-annual variability. Flexible options and strategies that allow for possible project mid-life adjustments as more information about climate becomes available (such as allowing for expansion of irrigated area in a wet climate) should also be considered.

Improving the Method

The methodology used in this book can be improved in a number of ways. For instance,

- Other climate scenarios could be included in the analysis.
- Simulated runoff could be better calibrated and validated with local data and by focusing specifically on dry years to get more accurate storage-yield curves.
- The impact of climate change on water requirements for both irrigated and rain-fed crops could be included.
- The seasonality of irrigation demand linked to the crop pattern could be used in calculating storage-yield curves.
- Evaporation at the dam and possible sedimentation over time have not been considered in the calculation but it could be included in the storage-yield curve calculation.
- The temporal scale used to build the storage-yield curve could be adapted to cover the main drought period; here, monthly data broadly capture the relative durations of dry and wet seasons, which both last several months; this approach could be refined where short periods of drought may be critical.
- Different crop patterns for wet and dry years could be included in the analysis to account for possible adaptation of the system.

Notes

1. Other project costs were US$6,500/ha (Kontagola, 11,000 ha under pressurized irrigation) and US$12,200/ha (Kano River Irrigation Project, gravity scheme, 18,000 ha).
2. This pattern reflects the potential of the zones, but the current crop production in public schemes can be different: maize is grown in the central and south due to uncertainty in irrigation supply.
3. http://mapspam.info/data/.
4. www.faostat.fao.org.
5. A more sophisticated approach would include a change in cropping pattern to make optimal use of the water available.
6. This assumption will be discussed in the results section.
7. There is no dam planned on the few sub-basins showing significantly drier climate in the future. Because the purpose of this study is to test the methodology—not to show

results from a particular site—dry climate characteristics (change in runoff and storage-yield curve) were applied to a planned dam on the sub-basin next to a dry climate sub-basin (Karma project).

8. Because the dam site is not always located at the outlet of the ArcSWAT sub-basin and because there is some bias in the simulated runoff due to the limited stations used or calibration, the mean annual runoff was used, as estimated in the JICA document as historical mean annual runoff and the changes simulated in ArcSWAT were applied to determine the future mean annual runoff for each scenario.

References

Block, P., and C. Brown. 2008. "Does Climate Matter? Evaluating the Effects of Climate Change on Future Ethiopian Hydropower." Paper presented at the Third Interagency Conference on Research in the Watersheds, Estes Park, CO, September 8–11. pubs.usgs.gov/sir/2009/5049/pdf/Block.pdf.

Defra (Department for Environment, Food and Rural Affairs). 2011. *Climate-Resilient Infrastructure: Preparing for a Changing Climate.* U.K.: Defra. http://www.defra.gov.uk/publications/files/climate-resilient-infrastructure-full.pdf.

Dessai, S., and R. Wilby. 2010. *How Can Developing Country Decision Makers Incorporate Uncertainty about Climate Risks into Existing Planning and Policymaking Processes?* World Resources Report, Washington, DC.

FAO (Food and Agriculture Organization). 2004. *Review of the Public Irrigation Sector in Nigeria.* Draft Status Report, EN PLAN Group, Rome. ftp://ftp.fao.org/AGL/AGLW/ROPISIN/ROPISINreport.pdf.

Hallegate, S. 2009. "Strategies to Adapt to an Uncertain Climate Change." *Global Environmental Change* 19: 240–47.

Hills, D., and A. Benett. 2010. *Framework for Developing Climate Change Adaptation Strategies and Action Plans for Agriculture in Western Australia.* Report of the Department of Food and Agriculture. http://www.agric.wa.gov.au/objtwr/imported_assets/content/lwe/cli/climatechangeframework_no%20cover_web.pdf.

Jeuland, M. 2010. "Economic Implications of Climate Change for Infrastructure Planning in Transboundary Water Systems: An Example from the Blue Nile." *Water Resources Research* 46: W11556. doi:10.1029/2010WR009428.

JICA (Japan International Cooperation Agency). 1995. *The Study on the National Water Master Plan. Sector Report Vol. 2.* Report prepared for the Federal Ministry of Water Resources and Rural Development, Abuja.

Kirshen, P. 2007. "Adaptation Options and Cost in Water Supply Report to the UNFCCC Secretariat Financial and Technical Support Division." http://unfccc.int/cooperation_and_support/financial_mechanism/financial_mechanism_gef/items/4054.php.

Lempert, R., and N. Kalra. 2011. "Managing Climate Risks in Developing Countries with Robust Decision Making." *World Resources Report 2010–2011,* Washington, DC. http://www.worldresourcesreport.org.

Ward, P. J., K. M. Strzepek, W. P. Pauw, L. M. Brander, G. A. Hughes, and J. C. J. H. Aerts. 2010. "Partial Costs of Global Climate Change Adaptation for the Supply of Raw Industrial and Municipal Water: A Methodology and Application." *Environmental Research Letters* 5: 044011. http://globalchange.mit.edu/files/document/MITJPSPGC_Reprint_10-13.pdf.

World Bank. 2007. *Africa Region: Irrigation Business Plan.* Washington, DC: World Bank. http://water.worldbank.org/water/publications/africa-region-irrigation-business-plan.

You, L., C. Ringler, G. Nelson, U. Wood-Sichra, R. Robertson, S. Wood, G. Zhe, T. Zhu, and Y. Sun. 2009. "Torrents and Trickles: Irrigation Spending Needs in Africa." Background Paper 9, African Infrastructure Country Diagnostic, World Bank Group, Washington, DC.

Environmental Benefits Statement

The World Bank is committed to reducing its environmental footprint. In support of this commitment, the Office of the Publisher leverages electronic publishing options and print-on-demand technology, which is located in regional hubs worldwide. Together, these initiatives enable print runs to be lowered and shipping distances decreased, resulting in reduced paper consumption, chemical use, greenhouse gas emissions, and waste.

The Office of the Publisher follows the recommended standards for paper use set by the Green Press Initiative. Whenever possible, books are printed on 50% to 100% postconsumer recycled paper, and at least 50% of the fiber in our book paper is either unbleached or bleached using Totally Chlorine Free (TCF), Processed Chlorine Free (PCF), or Enhanced Elemental Chlorine Free (EECF) processes.

More information about the Bank's environmental philosophy can be found at http://crinfo.worldbank.org/crinfo/environmental_responsibility/index.html.